Happy Birthday
1982

TALES UP!

Happy Birthday
1982

DEDICATION

To our many old "seat-of-the-pants" pilot friends who, over the years, have been the unwitting participants of just such high-larious aerial antics as are herein depicted; to the late international aerobatic champion, Tex Rankin, who taught us, often the hard way, to appreciate the humor in aviation—and to our Mother, Bertha Bohrer, who more often than not kept the vast majority of early-day pilots from starving to death, we most humbly dedicate this book.

TALES UP!

WALT & ANN BOHRER

*Foreword
by
Douglas "Wrong Way" Corrigan*

1971

Aero Publishers, Inc.

329 Aviation Road
Fallbrook, California 92028

COPYRIGHT ©

1971

AERO PUBLISHERS, INC.

All rights reserved.
This book, or parts thereof, must not be reproduced
in any form without the written permission
of the publisher.

LIBRARY OF CONGRESS CATALOG CARD NUMBER

77-157754

Cloth ISBN-0-8168-8800-0
Paper ISBN-0-8168-8804-3

PRINTED IN THE UNITED STATES OF AMERICA

FOREWORD

Since the stone age of aviation when hot air was first put to some semblance of good use as lofting power for balloons, airmen have shown a remarkable knack of getting themselves into ridiculous situations. This possibly explains why some early-day aeronauts were dubbed "balloonatics" rather than balloonists, or "aeronuts" instead of aeronauts. And while research has failed to turn up any more serious shenanigans by balloonists than dragging their anchors through Monday morning wash lines, or scaring farm wives out of ten years' growth by becoming entangled in fruit trees or windmills, the antics of the later airplane pilots have more than made up for this "loss!" From this point research has been productive!

Now the writer is not inferring that airplane pilots of the pioneering or barnstorming eras were a bunch of boobs. Far from it (after all, I was one of them)! They were, for the most part, a hard-working, often hand-to-mouth breed of guys who worked hard and long trying to prove in every way they knew that flying was safe and the airplane was here for keeps. It was just that motors in those days had a nasty habit of quitting, or konking out, at the most inopportune times. Instruments, too, were not exactly a pilot's dream. What pilot today would happily venture forth into the wild blue whatchamacallit with an oil gauge, a temperature gauge and a tachometer? Not very many! Aircraft radio was unheard of and navigational equipment for pilots was practically non-existent. If it had not been for the good old "iron compass"—the railroad tracks that ran from town to town, and the town names on the stations, half of America's early-day pilots would, in all likelihood, STILL be lost! And with the wood, wire and cloth "crates" of the times that could be set down on any old side hill cow pasture, generally atop one or more of a farmer's prize livestock, it behooves us to excuse most of the zany happenings experienced by "those daring young men in their flying machines!" There were, of course, those things that were neither the fault of plane OR pilot that resulted in some real weirdos, to snatch a modern-day phrase. But—and this is a BIG BUT—it was "pilot whims" that gave birth to the real aerial fiascos!

It was pilot whims that made a bored-stiff airline pilot, lumbering along between Los Angeles and San Francisco in a Ford tri-motor, loop his load of passengers near Santa Maria; that made a well-known airmail pilot dive at the crew of a dredge frozen in the Mississippi River and snag his wing on a guy line; that made a later well-known endurance flier plunk his ship down on a muddy stamp-sized mountain top for no reason at all—or an American soldier-of-fortune pilot pick a dog fight with a military plane over Nationalist China and force it down only to discover the plane was carrying none other than Chiang Kai-Shek!

Most of these things and many others are herein recorded for posterity by Walt and Ann Bohrer, a flying-writing team who, themselves, have been mixed up (in aviation, that is!) since the early '20's. Herein you will find things that happened to military fliers, bush pilots, airmail pilots, World War

I aces and just plain old barnstormers that shouldn't have happened to a dog—and there are several things in here that happened to dogs, too! Fortunately, most of these situations resulted in nothing worse than a few bent struts, some broken propellers, a few banged-up feelings and any number of very red faces. And you can bet a whole bucket of propeller pitch that most of the events depicted have been well-preserved under these old fly boys'—and gals'—helmets for a passel of years before being wheedled out of them by Walt and Ann!

I would like to call your attention to the fact that a number of these incidents were caused by pilots doing things the wrong way. Now why any pilot would ever do anything the wrong way is beyond me. In all of my years of flying, I have never been known to do anything the wrong way. This, naturally, includes several non-stop flights in my Curtiss Robin monoplane—most notably my 1938 non-stop flight through fog and darkness from New York to Los Angeles on which I landed in Dublin, Ireland due to a defective compass.

As I say, why any pilot would do anything the wrong way is a great mystery to me, but they do and some of the stories in the following pages prove it. But I've got to add that were it not for all these old boys who helped make aviation history the hard way over the years, Neil Armstrong, Ed Aldrin and Mike Collins would have had one heck of a time getting to the moon—and it's a good thing THEY didn't go the wrong way!

So just pull up the old rocking chair, fasten your safety belts and HANG ON! This will be quite a ride!

Douglas Corrigan

Douglas "Wrong Way" Corrigan
(Wrong Way to Ireland, 1938)

PREFACE

A number of years ago an Alaskan bush pilot at Cordova radioed for the weather in the Copper River district.

"What's the ceiling up there?" he asked.

He apparently got hold of a new ham radio operator whose knowledge of aviation could best be likened to Noah's knowledge of Apollo 12, for shortly this reply came crackling back through the earphones:

"I dunno for sure—but I think it's Celotex!"

Over Ohio one winter day pilot Cy Caldwell caught up with a slower moving aircraft—a Travel Air biplane. Since both were flying between two layers of clouds and the earth could not be seen, Caldwell concocted a fiendish idea. Coming up behind the slower craft so he could not be seen, Caldwell turned his ship upside down, passed the Travel Air and continued flying inverted for some distance. Looking back, Caldwell saw exactly what he had hoped for. The other pilot had sure enough turned his ship over and was now also flying upside down!

Up Yakima, Washington way, pilot Al McAllister was flying low over an alfalfa field one early morning watching several hunters out for pheasants.

Suddenly a covey of the birds zoomed up between his Waco biplane and the hunters below. Before he could bank off, four hunters opened up on them with shotguns. They missed their pheasants a country mile but every one of them scored a perfect bullseye on McAllister's aircraft! McAllister, thinking for sure the barrage had punctured his gas tank—as well as himself, high-tailed it back to the airport as though he were being chased by Satan himself; however, a ground check showed only a few score shot holes in the wings.

"It sounded as if I flew through a dad-gummed hailstorm!" declared McAllister trying to get his hair to lie flat again!

The foregoing incidents are typical of the nutty happenings experienced by airplane pilots since the day pioneer aeronaut Lyman Gilmore is alleged to have announced before one of his powered flights at Grass Valley, California, in 1905:

"I am not taking any chances—I am wearing my light fall suit!"

Since aviation's swaddling days unbelievable crazy things have happened to pilots, and newspapers seemingly have missed the boat by chronicling only aviation's tragic happenings. What stories they missed! What snafus went unrecorded! Accordingly, we have attempted herein to compensate for this gross lack by recording for posterity some of the more outstanding of these aerial "fox-passes." It is an ironclad cinch that the big fist of the Federal Aviation Administration (FAA), the agency that now, fortunately, governs aviation, has relegated such incidents as these to the ages of the past; yet they vividly illustrate the growing pains of a now vast industry and the unending wholesome humor of the men and women who, in their helmets, goggles and leather coats, helped bring aviation to the peak of efficiency it enjoys today.

And bear in mind, as you peruse these pages, the words of the little old lady as she watched a pilot repairing his motor after a forced landing in the hills of eastern Tennessee:

"Young feller," she cried, beaming upon the pilot, "if I was as smart in machinery as you be, I declare to goodness I'd go sommers and clerk in a hardware store!"

A DOUBLE DILLY BY DOOLITTLE and COLBY GOES INTO THE RED!

During one of the National Air Races performances at Cleveland in the early 1930's, General Jimmy Doolittle, then a lieutenant, was asked to fly over from Wright Field, Dayton, and put on one of his famous acrobatic shows.

Taking off from Wright in his Army Air Corps single-seater Curtiss Hawk fighter, he climbed to approximately 12,000 feet and was over the air races at the prescribed time of 2:00 p.m.

Pointing the nose of his biplane almost straight down, Doolittle headed directly at the Cleveland Airport with the thought of how he was going to "wow" the crowds right out of their seats.

Suddenly a wing snapped and peeled off his ship and Jimmy said aloud to himself:

"Doolittle, you have just lost a certain amount of your cunning!"

The finale of this aerial drama was that Jimmy bailed out in his 'chute and the Hawk piled up in a plowed field. No one hurt.

One of aviation's all-time greats, General Jimmy Doolittle.
In spite of his name, he's done a lot!

On another occasion, Doolittle was bringing a Boeing P-12 single-seater fighter in for a landing at Selfridge Field, Michigan.

Suddenly, over the end of the runway, the ship stalled out from under him and he wiped out the landing gear.

Upon filling out the accident report form, he came upon the question: What was the cause of the accident?

Thinking a moment, Doolittle put down his classic answer:

"Pilot suddenly ran out of experience!"

All of which points up the fact that, throughout the history of aviation, hundreds of pilots, either through loss of cunning, running out of experience, or for whatever reason, wound up in situations often more humorous than tragic, and that it can happen to the best of them!

It would be difficult, for instance, not to imagine a flying paint salesman's airplane upside down on the ground without a bucket of red paint over his head.

Tom Colby, who for years flew a vividly varicolored Waco hither and thither as representative of the Berry Brothers Company, manufacturer of paints and enamels, found himself in just such a "cartoonist's dream" one somewhat-less-than-bright Minnesota day.

But let's let Colby, now a citizen of Pauma Valley, California, tell it like it was:

"In the summer of 1934 I was on a flying business trip from Detroit, Michigan to Winnipeg, Manitoba. The plane was an early type cabin Waco. It was lightly loaded with only myself, personal luggage and a gallon of bright red Berryloid Quick Drying Enamel.

"Over Detroit Lakes, Minnesota, I spied a small airport and thought it prudent to land for some badly-needed fuel. No one was on the field but I hailed a gas trucker going by and was assured he could fill me up with aviation gas. It must have been lightly 'watered' as the motor sputtered and finally quit at a critical moment after take-off. With nothing to do but come down, the Waco flopped to a rapid rest upside down in a wet and boggy pasture. The can of red paint popped open and yours truly was well drenched over his bare head and shoulders with the fast-drying contents. Tumbling rapidly out, I stood up to view the damage and saw two horrified farmers staring at the 'bloody' sight. It must have been just too much for their queezy stomachs and away they ran!

"Later they were convinced that I really wasn't cut to pieces and gave all needed assistance. A plentiful supply of alcohol and turpentine mixed up by the local druggist was the only medicine required and a rough scrubbing ensued, taking some skin and plenty of hair with it.

"At the time the incident wasn't so funny but in after years I've had many a laugh over those horrified country guys who couldn't stand the sight of 'blood.'"

Here's that flying paint salesman, Tom Colby, without a can of red paint on his dome! The golfy-looking fellow with him is Lee Schoenhair (right) whom many will recall as a well known racing pilot. The ship is a Buhl Air Sedan flown by Colby and Schoenhair in the New York to L. A. Air Derby in 1928.

WILEY POST TAKES A DIVE!

While the authors of this book had the rare pleasure of flying with Wiley Post on a "snitched" barnstorming junket for several days about a year before he and Harold Gatty made their record breaking world flight, nothing overly humorous happened, although Post was excellent material for something TO happen to!

We say a "snitched" barnstorming tour because, at the time, Post was supposed to be on a demonstration tour for Lockheed showing off their high-speed Vega monoplane. However, at Portland, Oregon, he decided to take time out to "go and carry a few passengers at Cannon Beach, Oregon, and Long Beach, Washington," both popular summer meccas. We were invited to go along as his "crew" since we knew the area, having barnstormed there before with Tex Rankin.

While at Cannon Beach, a small kid rode up on a bicycle and Post asked him if he could try out the bike. The kid, sharp for a ten-year old, said he sure could if he'd reciprocate with an airplane ride. It was a deal. But, 45 minutes later, with at least twenty rides sold and waiting, Post and the bike were nowhere in sight. At last, in about an hour, here he came peddling back. He'd been clear to what is known as Hug Point, about six miles down the beach, and was so pooped and stiff he could hardly crawl into the airplane.

"That's the last time ah'll ever ride a bike," he moaned, "mah seat's so sore ah kin hardly sit on it!"

However one thing did happen to Wiley a few days before he and Will Rogers had their unfortunate crash.

He and Rogers had arrived in Dawson, Yukon, and as all fliers usually wound up, were as hungry as bears. There they met Harriet Malstrom, a gal writer-photographer from Seattle who had flown to Dawson for a story. She was hungry, too, so the three of them had a big moose steak dinner at the Arcade Cafe.

During the meal the discussion centered on the route to be flown to Barrow.

"You are worried about the route to Barrow," put in Will, "but what I'm worried about is how low you will have to fly to get under the Arctic Circle!"

Wiley was still stewing about the route after the moose steak repast, so he and Miss Malstrom decided to hike down to their float plane, anchored in the Yukon River, and get his maps, a decision that was to make the famous Yukon even more famous.

Hurriedly approaching the red and silver Lockheed, Post made a flying leap over to the nearest pontoon. It was raining and the pontoon was as slippery as a mossy rock from a mixture of engine oil and rain. The result was that Post lost his equilibrium and, with his arms and legs flailing to the four winds, went kersplash right on his back into the swift, glacier-chilled Yukon.

Up he came blowing and snorting like a whale, with a dazed and thoroughly bewildered expression on his face. But when he saw Miss Malstrom doubled up laughing on the dock, he managed to get that big Oklahoma grin going again.

Later that evening when Miss Malstrom went over to the Royal Alexandria Hotel, Post's and Roger's Dawson headquarters, there were his clothes, from his "undies" out, hanging up to dry in the main hallway of the hotel with half the town in there looking at them!

Said Rogers to Miss Malstrom: "It was all your fault, young lady. If you hadn't been distractin' his attention so much he'd have watched where he was goin' instead of fallin' into the river. That's the way with you women; you're always jabberin'!"

Wiley Post's last picture, taken by Harriet Malstrom the day after Post had fallen into the Yukon, and the day before his final flight.

A CASE OF "LEND ME YOUR EARS!"

Have you heard about the time Amelia Earhart, W.B. Kinner and Neta Snook "snuk" the corn—and were caught with their pants loaded?

This tale stems from the lean days of aviation when most people would "fly" only if they could "keep one foot on the ground" and aviators were really hungry!

Everyone has heard of Amelia Earhart and most people who have followed aviation at all know that W.B. Kinner was the manufacturer of the famous aircraft engine bearing his name and, earlier, designer-manufacturer of the "Airster" biplane. Not too many may recall Neta Snook, however, although Neta was quite a name in early-day aviation.

In 1920 Miss Snook (now Mrs. W.I. Southern of Los Gatos, California) was the only female operating a commercial flying field in Southern California—or perhaps in all the West. This was a 50-acre plot located on the west side of Long Beach Boulevard at Tweedy Road about half way between the cities of Huntington Park and South Gate. This field, on which vegetables had previously been grown, had been leased by Kinner and was known as Kinner Field. On it Kinner had built a wooden hangar and installed a gas pump.

It was in this hangar that Kinner designed and built his first "Airster," a small 17-foot-wingspan biplane which was to become the nucleus of his dream of "a small plane for every household." This same little "Airster," powered by a three-cylinder Laurence engine, later was purchased by Amelia Earhart—her first airplane!

Pilotesses Nita Snook (left) and Amelia Earhart in front of 3-cylinder Kinner Airster near Long Beach, Calif. in 1921. Amelia had long hair then. Later her mother allowed her to cut it!

14

Since Kinner himself was not a pilot, he had turned all commercial flying rights on the field over to Miss Snook in return for her test-flying the first "Airster" and other "Airsters" to come.

Miss Snook had been flying since 1917 and had been in the service of the British Air Ministry during World War I. After the signing of the Armistice in 1918, Miss Snook purchased a wrecked Canadian training plane, popularly dubbed a "Canuck" or Canadian "Jenny." This ship was a contemporary of the American Curtiss JN4 ("Jenny") used to train American pilots during World War I. The only difference was a slight variation in the shape of the tail section. With an eye to the future, Miss Snook shipped her wrecked "Canuck" to Ames, Iowa, where she repaired and assembled it in the back yard of her parents' home. That summer she flew at fairs and carried passengers throughout the Midwest. With the coming of fall and cold weather, she took her ship to California where Kinner Field became her base. It was in this rebuilt "Canuck" that Neta Snook gave Amelia Earhart her first flying lessons.

But back to the corn!

As has been mentioned, those were the lean days of aviation and both Amelia and Neta—and, more often than not, Mr. Kinner—were quite hungry at times. Cash on hand was a rare commodity and what little did come their way was used for gasoline and spare airplane parts. However, adjacent to Kinner Field a most convenient field of succulent corn happened to be growing. The opportunity to combine these tasty ears with a kettle of boiling water in the Kinner hangar, together with a bit of butter, was a treat not

to be easily overlooked! The result was almost daily forays into the cornfield by the Misses Snook and Earhart and, of course, the equally hungry Mr. Kinner!

One bright corn-season day, four or five rows into the "goodies," this hapless trio came face to face with the irate owner of the cornfield.

"What," bellowed he, "are you doing in my cornfield?" bringing vividly to their minds the story of "The Three Bears" ("Who's been eating my porridge?").

Naturally, it was hardly possible for the trapped trio to say they were

15

searching for a missing propeller, especially with ears of corn protruding from every pocket and bulging the length and breadth of their coveralls. Somehow the farmer didn't quite subscribe to the common premise of the time that an aviator could do no wrong, and attempting to explain their position was utterly useless with ears of corn now spilling out of the stuffed legs of their coveralls. However, the speed of flying youth has ever won over the age of contrary farmers. The three were practically airborne getting out of that corn patch and corn was again enjoyed in the Kinner "kitchen"—albeit for the last time!

During this period also, Miss Snook was approached by a group of pilots organizing a barnstorming tour of the Orient. The group, known as "Barr's Flying Circus," was using a strip of beach at Santa Monica as a flying field and asked Miss Snook to join them. Her part was to perform such acrobatics as loops, rolls and spins, and to fly one of the two planes used in a change-plane act. The latter maneuver required practice and for this Miss Snook would fly to their field or they would come to hers. The wing-walker would climb to the outer tip of the wing of one plane and the second would fly close enough for him to grasp the wing-skid and climb aboard.

One morning Miss Snook flew to Santa Monica for a practice session. There Mr. Barr, head of the group, pointed to a newly overhauled Jenny and told her to take it up a couple of thousand feet and do a few loops. The first thing she found when she climbed into the cockpit was that she could barely reach the rudder bar. To remedy this they took a cushion from another plane and placed it behind her back. Upon taking off she found the plane was right-wing heavy, meaning that it was not properly rigged—or aligned, so instead of attempting to loop at 2,000 feet, she climbed to a much safer 5,000, a wise move as you will see.

At 5,000 feet Miss Snook pulled into her first loop. She stalled at the top and the cushion behind her back fell out. As the plane began falling upside down, the engine oil began streaming out of the breather pipe, covering her head and goggles with warm, gooey oil—an unladylike decoration even for an aviatrix! Then the ship went into a spin—or tailspin as the maneuver was known in those days. There she was, dangling upside down, not able to see because of the oil and barely able to touch the rudder bar because of losing the cushion. This was a neat pickle for this gal to be in, especially since hard reverse rudder is necessary to come out of a spin. Stretching her leg with all her might, she pushed the rudder bar as far as she could with her toe. After a seeming indeterminable age, the plane began to right itself. As soon as the oil stopped pouring out, she tore off her goggles and found herself looking straight down into the ocean with the altimeter showing 500 feet—barely enough altitude for a quarter turn! After a shaky, but somehow satisfactory landing, she taxied up to the line, expecting to be congratulated for being cagey enough to climb to 5,000 instead of 2,000 feet, thereby saving herself and the plane. Instead, she was greeted with an angry:

"What's the big idea of losing our best seat cushion in the ocean?"

Amelia's father, Edwin Earhart, subsequently talked her out of becoming a member of the Barr Flying Circus, probably for fear that daughter, Amelia, would follow suit!

NO BOUQUETS FOR WILSON!

A beautiful summer evening and nothing to do is a combination well capable of spelling trouble with a capital "T" for an idle pilot.

Just such a combination befell pilot Claude Wilson back in the latter 1920's when he was manager of the Salinas, California, airport. "It was one of those things that shouldn't have happened to a dog—not even an Airedale or a Skye terrier!" said Wilson as he reminisced over the events of that long-back evening.

Being one of those rare fogless, windless and balmy evenings with which the Salinas area is not too often blessed during the summer, Wilson decided it was a perfect occasion to drop a box of candy to the two daughters of a friend in the Blanco district, a farming area some 10 miles west of town. He thereupon dispatched one of his flying students into town with some money

This dapper appearing plus-four attired pilot being held up by his Curtiss Robin monoplane is Claude Wilson of Aptos, Calif. An expert on how to pick sweet peas from an airplane!

17

to buy the candy while he fashioned a parachute with which to safely drop the "goodies" from the airplane.

Upon the student's return from town Wilson hopped into his airplane, an open-cockpit Eaglerock biplane, and headed for the farm of his friend, a gent who specialized in growing prize sweet peas for show purposes at various fairs throughout the state—flowers which had consistently won him first prizes wherever they were exhibited.

Arriving over his friend's farm, Wilson circled the house, gunning his OX-5 engine a few times in order to get the family out into the front yard. At precisely the right moment he throttled back, leaned out and dropped the box of candy attached to the parachute. The miniature 'chute opened and the candy made a perfect landing directly in front of the waving family.

Immensely pleased with himself and his "bombing" prowess, Wilson pushed open his throttle to head back to the airport. At that precise instant the motor quit dead! Having lost much of his flying speed by gliding down around the house, there was nowhere to go but straight ahead—right into the middle of his friend's prize sweet peas, into which he plopped, mowing a good half acre of the finest specimens flatter than a pancake!

To have been shoveling coal in the utmost retreats of Hades would have been sheer pleasure in comparison to his feelings as he sat in his airplane dejectedly awaiting the arrival of the farmer and his family.

His farm friends graciously passed off the complete ruination of their prize flowers as merely "one of those things" as Wilson shamefacedly repaired his engine.

Of course they did not take into account that Wilson would mow down another half acre of sweet peas on his take-off—which he did!

On subsequent balmy summer evenings Wilson was content to sit under the wing of his airplane and twiddle his thumbs, leaving well enough alone!

INTERNATIONAL INCIDENT!

An adage concocted by an old pilot friend of ours during our early barnstorming days: "Give a pilot enough rope and he'll try to pull something!", stands us in good stead in relating the following incident involving several well-known early-day fliers among whom were Navy Lieutenant (now Admiral) "Put" Storres; Lieutenants "Woody" Woodring and W.L. Cornelius of the then Army Air Corps; one Bob Lautt and the guy who told us this story—Vince Barnett.

Many of our readers will recall that Lieutenant Storres was a member of the Navy's crack aerobatic team, "The Three Sea Hawks," forerunner of the current Navy "Blue Angels," and that Lieutenants Woodring and Cornelius were members of the Air Corps' famed "Three Musketeers," the third member being Lieutenant (now General) Jimmy Doolittle. As for Vince, most of our readers will recognize him as a noted motion picture character actor and Hollywood's famous "ribbing" emcee. He's also a pilot.

This "intrepid" looking fellow is Vince Barnett, famed ribber, actor and pilot. Although looking as if he'd just bitten into a green persimmon, he's actually thinking of a bus ride he once took.

19

Now you turn fellows like these loose on a town and ply them with liquor and something is bound to "give"—only in this case it wasn't "give"; it was "take!"

The fact that the "Sea Hawks"—the other members of which were Lieutenants D.W. "Tommy" Tomlinson and William Davis (now Admiral Davis)—had just "creamed" the "Three Musketeers" at the 1928 National Air Races at Los Angeles, made not one whit of difference. It simply called for a party and the logical place for the type of party called for was Tia Juana, just over the Cal border in Mexico. The logic here, if any, was that prohibition was the rule in the states while booze flowed freely and legally "Souse of the Border," as Barnett quite aptly puts it.

The party went fine and all concerned managed to get out of Tia Juana without being incarcerated in the local calaboose, a feat in itself. It was at the international boundary that their trouble began.

Alert immigration officers (and they didn't have to be too alert) decided the boys were too far in their cups to drive a car and, accordingly, ordered them to park and take a bus standing nearby. So into said bus they piled whereupon Storres immediately turned on the ignition, started the motor and headed north. After a couple of "pit" stops, the merry crew wound up at the Naval Air Station in Coronado.

The police, called by the busless bus crew who were coffeeing up when their steed was purloined, finally found the bus and rounded up its late occupants. All were taken before the night magistrate for their come-uppance.

In their defense, "Put" simply stated: "They told us to park our car and take the bus, so that's what we did!"

When the arresting officer explained that the bus was legally parked in front of Storres' quarters, they were duly released with the court's admonition to go and consume plenty of black coffee!

FORSSTROM'S FLYING FOLLIES!

Probably one of the better known pilot-instructors on the Portland, Oregon scene during the 1930's was Carl Forsstrom, a sharp pilot with an equally sharp Swedish sense of humor. In this heyday as a pilot, Forsstrom managed to accomplish a few "hilarious heroics" in the field of flying, not the least of which was the time he caused the supersonic evacuation of a hangar in which an evening ground school class was in progress. Thinking he'd have a little fun with the ground school students, Forsstrom pointed the nose of his Fleet training plane down at the roof of the hangar-class room and pushed his throttle wide open. About a 100 feet from the roof, he decided he'd better pull out and hauled back on the stick lest he joined the class, airplane and all. But the Fleet had accumulated so much downward momentum that, with its nose now up, it continued on down until the landing gear hit the roof with a resounding whack that must have been heard for 30 miles, and one can well imagine what it sounded like inside of that tin hangar. The result was that no school fire drill had ever evacuated a building as fast as that hangar emptied! Seventeen students were outside in less than ten seconds, a record that could never have been established had they stopped to open doors and windows. Meanwhile Forsstrom's Fleet had bounced off the roof across a main highway and had somehow regained flying speed.

Coming in for a landing and taxiing up to the line, Forsstrom innocently asked, "What's everybody standing out here for?" as he slowly felt over his head for a halo!

The ground school class discovered then why he had been nicknamed "Buzz" Forsstrom while a cadet at March Field!

However, the day Forsstrom accidentally arrived in the interior of the Portland Airport Administration Building in an American Eaglet monoplane is a situation far more worth the telling, especially since no one was injured other than being well-dusted and adorned with various and sundry debris. And, from this point, we turn the story over to Forsstrom himself:

"Flying wasn't nearly as complicated in 1931 as it is now. About the only thing that hasn't changed down through the years is the law of gravity which, of course, is what flying is all about.

"Flying was a lot funnier in the earlier years though and, because of our ignorance, considerably more dangerous. I don't mean to say we were stupid—at least not all of us; there simply were a lot of things about aerodynamics that we hadn't found out yet. We were smart—but ignorant.

"For instance, we had never heard of a high-speed stall. If someone had told us you could stall at high speeds, we would have laughed him right out of his OX5 Waco 9. We wouldn't have worried anyway because we didn't have any planes with high speeds. A guy who said he had flown the 120 miles from Portland, Oregon to Eugene, Oregon in an hour either was a darn liar or he had a 30-mile tail wind.

"Our instruments consisted of a tachometer that usually worked, a clock that didn't, an oil pressure gauge and an altimeter. Sophisticated panels sported a turn and bank indicator which was a danger in itself as it led some of the boys to experiments in blind flying. There were a few casualties among these brethren, caused usually by groping around in the 'soup,' too close to the mountains. In Northwestern Oregon you are always, 'too close to the mountains.' This type of mishap was referred to locally as 'finding a cloud with a rock in it.' The experimenter who survived the experiments reported they usually came out of the overcast upside down, in a screaming dive, or both.

"One of the private schools on Portland's late Swan Island Airport installed a hood over the rear cockpit of a Menasco-powered Fleet for blind flying instruction. They offered me a courtesy flight which I accepted readily. However, when they locked me under the hood, I immediately contracted a violent case of claustrophobia. I cured it by beating on the inside of the hood with both fists and yelling, 'Let me out of here!'

"They promptly rescinded the invitation they had given me and to this day I am unable to fly blind. Only last week I ran into a partly open door in the middle of the night—on my way to the bathroom!

"In the early 30's there appeared on the scene a rash of what today would be called 'mini-planes!' They had several things in common, most notable of which was a severe shortage in the horsepower department. They compensated somewhat for this by using a high-lift wing and keeping all other weight to a minimum which gave the pilot the feeling that he was flying a wicker basket. Oddly enough, these contraptions were able to gain

Down at March Field they dubbed him "Buzz".
We wonder why anyone would hang a sobriqet
like that on Carl Forsstrom?

considerable altitude, even with two people aboard. This gave them a built-in danger factor similar to that in a tall stepladder.

"I came down to the field one Saturday morning and found one of these small monsters warming up in front of our hangar.

" 'What's that out-size electric fan running out there for?' I asked our part-time mechanic who was also one of my students.

" 'The boss said to tell you to check me out in it,' he replied. 'It's owner owes him some time and this is how he gets it back.'

" 'I'll tell you what,' I told him. 'Let's run up and down the runway, flapping our arms. I'll show you how and you'll get more out of it than flying that bird cage!'

" 'No dice,' he said. 'I've got some solo time coming and I want it!'

"I went out to look at our conveyance and what I saw didn't improve my appetite for the 'Wild Blue Yonder.' It was equipped with a three-cylinder radial engine of uncertain horsepower and a certain reputation for popping cylinder heads clear off and peppering the countryside with pistons. Fortunately, it only had three. Also the tachometer needle was circling the dial completely, at about 50 times a minute!

"The owner of this disaster was having a hamburger in the airport cafe and I went over and asked him what kind of a reduction factor he used in reading his tachometer. He snorted into his coffee cup. 'Don't look at it while you're flying,' he warned. 'It'll make you dizzy as a bed bug. Just shove the throttle against the fire wall and leave it there.'

" 'How about taxiing?' I asked with as much sarcasm as I could muster.

" 'Same thing,' he said, 'wide open.'

"I went back to the apron and put on my helmet, goggles and six-foot silk scarf—remember, this happened 38 years ago—and climbed into the rear seat of the 'Whatever' to run her up. I pushed the throttle slowly to the fire wall. Nothing happened for several seconds and then there seemed to be a gradual increase in noise from somewhere up front. The tachometer needle kept its steady sweep of the dial at 50 rpm and I stared at it with the uneasy feeling that it was turning faster than the engine. I started getting dizzy and wrenched my eyes away from the tach with an effort.

"The noise up front seemed to be getting louder and when it began to sound like a bushel of walnuts pouring into an empty steel drum, I throttled back.

" 'Kick the chocks out,' I yelled at my student. 'We'll never be able to fly this thing over them.'

"He did so and climbed in the front seat. 'Taxi way down to the end of the island,' I told him. 'We're going to need all the room there is.'

"Swan Island had an accidently built-in safety factor for take-offs. There were no trees or barriers of any kind at either end—just an abrupt drop of from five to fifteen feet, depending on the height of the Willamette River. Aircraft that didn't want to fly could sometimes be coaxed to by taxiing them at full speed off the end of the island where they automatically became airborne at the end of the runway.

"Usually they gained confidence when they found themselves in the air and this, coupled with the 14-foot dive, gave them air speed. I point out that if the river was up to 12 feet, you could dive only 11 feet. The higher the river, the shorter the chances.

"Prospective students were sometimes startled when we asked them causally if they could swim. Our reason for asking the question was sound.

"I mention these facts to make it clearer why I was having the student taxi so far up stream. Actually, I was thinking if we had floats along with the wheels, I would make him go clear to Oregon City. As it was, it took so long to taxi to the end of the runway that I began to worry that we'd have to go back for more gas.

"We finally made it, turned on to the runway and the student started the take-off. After trundling along with the tail down for the first hundred yards, I got the impression that we had made better time on the trip down from the hangar. Unfortunately, the tail finally came up and after scattering gravel for another few rods, we became airborne. The owner was right about handling the throttle. There couldn't have been more than 10 mph between the taxiing speed and flying!

"At about 30 feet altitude, which apparently was close to this Gismo's service ceiling, the right wing began to drop. 'Get that wing up!' I yelled. He couldn't seem to do it. I shook him off the stick and I couldn't seem to do it either. The airplane ignored us, steepened the turn a little and lined us up neatly with the administration building. 'Cut the switch and cover up!' I shouted. He did both in one motion.

"I kicked top rudder, crossed the controls and prayed that I could stick the right wing in the ground before we both got administrated. No results. At the last possible second I gave up trying and buried my head in my arms. There followed a rending crash, a couple of bumps, the sound of breaking glass—then silence. We looked up cautiously and found we were sitting in an airplane fuselage inside the lobby of the administration building, which also served as a passenger waiting room!

"The airport superintendent was standing there by a desk with a handful of papers. He didn't say anything. He just stared at us. We didn't say anything either. We just sat there and stared back at him. I finally felt the tension rising, so I cleared my throat and said brightly, 'Hi there!'

"At the investigation they said there were several reasons why we didn't get killed but having good sense wasn't one of them.

"They cited what they called 'decelerating factors.' They seemed to be proud of the phrase because they kept repeating it. I finally figured out what they meant was when the landing gear hooked onto the balcony fence, it slowed us up some. The right wing being knocked off by the steps and the left wing getting it from the archway slowed us up some more. The fuselage, missing a solid steel support by some six inches, was a bonus. By this time all we were riding in was the fuselage and when it went through the plate glass door, we were decelerated to a standstill. I've had a lot rougher stops in California traffic.

"They said it would cost $500.00 to fix the building, I agreed that they had quite a hole in their property but I said it wasn't as big as the hole in my pocket. Besides that, I told them, if a barnstorming pilot had $500.00, he wouldn't even know where to put the decimal!

"I'll say one thing for them—they were patient with me. They waited 25 years and then finally knocked the rest of the building down with one of those big iron balls on a string. I wish I could have been there to see what the heck that building was made of. Sponge rubber, I'll bet.

"One day not long afterwards a guy came along and offered real money (not very much, but real) to go to work as a traveling salesman. I gave up flying and started eating every day.

"I've had a few close calls in this business too, but for different reasons. I guess trouble is where you find it, in the air or on land. If this doesn't work out right, I may go to sea!"

BAIL, BOWMAN, BAIL!

When an earthquake in Japan causes a tidal wave on the Pacific Coast of North America, that is news. But when such an event causes a couple of fliers to spend hours bailing a goodly portion of the Pacific out of their aircraft's hull off the California coast, using their shoes as bailing buckets, that is ridiculous!

But such is exactly what happened to Ray P. Bowman, well known Oakland airman, and a news service representative back in 1931 after one of Nippon's more severe shake-ups.

It seems the International News Service (INS) was quite anxious to get its quake photos in the U.S. news media first. So they made arrangements with the Curtiss Wright Company, for which Bowman was Northern California distributor, to fly their representative out beyond the Faralon Islands to meet a trans-Pacific liner from which the pictures would be tossed overboard to them. They would then take off and deliver said photos to San Francisco, thus scooping the competitive news services.

Had Ray Bowman, far right, in this 1928 photo snapped at Oakland, known he was going to have to bail half the Pacific out of his airplane, he'd have bought boots instead of shoes! His flying buddies, from left, are Jerry Andrews, Denny Wright, Capt. Bill Royle, Louis Thaden, Swede Leimington, Capt. Bill Fillmore, Swede Anderson, Major Bernie Foster and Nolan Bannister

Using a Loening Amphibian commuter, the two flew out to the prescribed spot. Just as the landing at sea was made, a huge swell hit the bow of the amphibian and loosened quite a number of rivets. This was not detected for some time, until water appeared above the floor boards. By then the additional weight of the water made it impossible to take off. To make bad matters worse, there was no bilge pump on board—usually imperative equipment on all seaplanes or amphibians. All they had to bail out the fast incoming sea with were their shoes and, says Bowman, it was a good thing both had big feet! In the meantime, adding insult to injury, the trans-Pacific ship they were to meet, passed by, paying not one whit of attention to the two fliers. No pictures were dropped and the ship sailed merrily on.

It took over five hours of hard bailing with their shoes for Bowman and his INS cohort to reduce the water enough to permit a take-off which was finally accomplished.

Needless to say, the report filed by that particular INS representative was turned down by the telegraph office due to his ungentlemanly choice of words.

ANYTHING TO OBLIGE!

From the foregoing episode we are reminded of another flying instance involving the Fourth Estate—or, for those not in the "know," the press.

An airmail plane had plowed into a mountainside deep in the Cascade Mountains of western Washington and a couple of Portland, Oregon, newspapers had contracted with Portland's famous pilot, Tex Rankin, to fly them over the crash site for pictures.

Just as Rankin was nearing the crashed plane, the engine of his Stinson monoplane commenced to sputter. Tex immediately turned and headed out of the mountains toward the Columbia River Gorge where he stood a chance of finding a pasture or a sand bar on which to make a forced landing.

Meanwhile the two photographers had been so busy getting their camera gear ready for action on the mail plane crackup that they missed hearing the engine sputter, nor did they note the plane turning back. One finally noticed Rankin heading out of the mountains and away from the crash.

"Hey, Mr. Rankin," he cried, "aren't we getting kind of far away for a good shot of that wreck?"

"Well," answered Tex, "if you can wait a few minutes, we'll have a wreck of our own to shoot!"

DID ANYONE HERE SEE JENSEN?

If the good citizens of Petersburg, Virginia, ever complained that "nothing ever happens in this town," we can jolly well assure them something DID happen, only the whole kit and kaboodle of them apparently slept through the entire unbelievable event.

In 1928 on the occasion of one of heavyweight champion Gene Tunney's major championship fights in Miami, most metropolitan New York area newspapers were out to scoop one another with the fight pictures. Since this was before the days of wirephotos, most of the dailies had chartered airplanes to fly to Miami and bring back the pictures. This developed into a wild race with some 40 airplanes from the New York area competing, including some sent by questionable outside interests to fly back motion pictures of the fight. The fact that it was illegal at that time to fly fight movies from one state to another—a law that has long since been repealed—apparently did not deter these pilots in the least. To them a buck was a buck.

Martin Jensen, second-place winner of the famous 1927 Dole trans-Pacific air race from Oakland to Honolulu, had been doing some flying for the New York Daily News and so was commissioned by that paper to fly to Raleigh, North Carolina, where he was to meet a pilot flying north from Miami with the pictures.

Flying his famous Dole trans-Pacific Breese monoplane, "Aloha," Jensen hopped down to Raleigh for his relay rendezvous. Arriving at that city in the early p.m., Jensen all but wore a groove in the field pacing back and forth awaiting the Miami pilot. In true Pony Express-style, he had left his engine running to be able to effect a quick take-off when the northbound

Martin Jensen in his Breese monoplane "Aloha" taking off from Oakland for Honolulu in the famous 1927 Dole Race. Jensen placed second, the race being won by Art Goebel.

pilot arrived. But as the sun sank lower and lower toward the horizon, it was apparent the Miami pilot had been delayed. Finally, at sunset, he zoomed in, the pictures were transferred and Jensen took off for Newark in the final throes of daylight.

Since the "Aloha" was a cabin monoplane with an open pilot's cockpit immediately aft of the engine and a passenger cabin behind the cockpit, helmet and goggles were necessary gear for the pilot. Jensen always carried three pairs of goggles—light, dark and amber; the dark pair for flying in bright sunlight and so on.

As darkness fell about 30 minutes after his take-off from Raleigh, Jensen suddenly realized he was still wearing his dark goggles, with the light and amber pairs in his briefcase back in the cabin. He couldn't turn back to Raleigh as that would have defeated what he had set out to do for the News—scoop the rival papers. At the same time overcast conditions made it darker than it ordinarily would have been at this time of the evening. All in all he couldn't see too well. Shoving the dark goggles up over his helmet he tried flying without them but the prop wash made his eyes water.

Arriving over Petersburg, Virginia, about 9:30, Jensen circled the town but could locate no landing field. Nevertheless he was bound and determined to get his light goggles. Circling once more he picked out a so-so lighted side street and came down, landing between wires, ditches and over a high-sided

A happy Martin Jensen and his
pre-Dole Flight Jenny.

culvert through which the "Aloha's" tail barely fit, but down he was! Taxiing to a stop, Jensen quickly retrieved his light goggles from the briefcase, hopped back up into the cockpit, opened the throttle and took off, again dodging wires, poles and street lights. And as the first automobiles came barreling down on the scene, Jensen zoomed over them at 100 feet, leaving them nothing for their curiosity but a swirl of dust and other take-off debris. Somehow he safely cleared Petersburg, though to this day he still wonders how, and proceeded on to Newark.

By this time, he was so low on gas he decided to make a refueling stop at Richmond, landing there about 10:30 p.m. Meanwhile a large storm front had moved into the area and the weather was something far less than a pilot's dream. Consequently, when Jensen asked to be refueled, he was emphatically told that he could not take off that night—everyone was grounded.

"Oh, I just want to gas up tonight to be ready for an early take-off in the morning," was Jensen's reply. This seemed to satisfy the airport attendant.

"Now, tell me where I can park my airplane?" asked Jensen after the ship had been refueled and the bill paid.

The attendant pointed out a spot and Jensen headed for it. Suddenly he turned, gave it the gun and took off, heading off into the storm for home!

The weather was so bad that all he could see of Washington was a faint glimmer of light through the muck, but he made it to Newark, arriving there about midnight. The Daily News car was waiting and the paper scooped all others with photos of the fight by many hours; no other picture plane arriving 'til the following day.

Jensen's aerial espisode had an anti-climax six months later when two husky U.S. Marshalls knocked on the door of his New York apartment.

"Your name Jensen?" he was asked.

"That's me," replied Jensen.

"Well, you'll have to come with us. We want to ask you some questions," he was told.

Taken to the District Attorney's office, they wanted to know who sent him after the fight pictures. Jensen wouldn't talk. After being grilled for three hours, the D.A. said, "Throw him in the cooler!" Right then did our hero become as talkative as a minah bird and told them the New York Daily News had sent him.

"Well, why didn't you tell us that before?" they asked. "We only want the ones that carried MOVIES from one state to another!"

"That was my narrowest squeak from acquiring a striped sun tan!" says Jensen with a hearty ha ha!

IT WAS ALL DOWN HILL – – BUT . . . !

What happens when the whimsey of a pilot causes an airplane to be plunked down on a mountain top for no reason at all, except to merely see if he could do it? We'll let Karl Voelter of Coral Gables, Florida, well known ex-Marine Corps, CAA/FAA and Curtiss Wright pilot who has passed his 50th flying anniversary, answer this one in his own way:

"In the winter of 1929 we received in our New York office a request to demonstrate a Curtiss OX-5 Robin to the Goodyear Rubber Company in Akron, Ohio. At that time I was Sales Promotion Manager for Curtiss-Wright Flying Service. That was the sales and service organization for the parent company. True, we had a branch office in Cleveland which ordinarily should have handled the matter. But for some reason, our manager at that location—Louis Meister, thought the deal should be followed by headquarters. Accordingly, it fell my lot to arrange the trip and subsequent demonstration. I decided to take the assignment personally. I planned to leave Roosevelt Field the following Sunday.

"On Saturday, however, one of our best demonstration pilots, Dale "Red" Jackson, returned from a field trip and came into my office, as was his usual custom. In the ensuing conversation I mentioned the Cleveland project and asked him if he'd like to go along. Always enthusiastic, "Red" was quick to reply in the affirmative.

"Early the following morning we met at the CWFS Sales Hangar at Roosevelt where our airplane, a shiny new Robin, was ready. It was February and cold that morning. We both wore heavy overcoats and, singularly, both of us wore derby hats. We flipped a coin to determine who would fly the first lap of our trip which was to be to Syracuse. I won the toss so chose to ride, rather than fly, principally in order that I could read my morning paper in hopeful leisure.

"Our flight course was across the mountains rather than the usual long route via Albany. As we approached the hills, the terrain was rugged but beautiful in the morning sunlight. An ice storm, probably the night before, had left the trees and bushes festooned and sparkling. Things were apparently going smoothly. The air was smooth. Visibility was almost unlimited. "Red" was courteously quiet and I was absorbed in my paper. Occasionally, as I glanced out the windows, it was apparent that "Red" was flying his usual pattern—just skimming the trees and knolls. He had an inbred impression that regardless of what might happen, he could get a Robin down safely.

"Suddenly, and without warning, "Red" hollered, 'Hook your belt, we're landing!' I grabbed for the belt and as I glanced out the window I could see that he was approaching a little clearing atop one of the mountains. Almost immediately we rolled onto the apparently frozen ground and just as quickly we discovered that the heat of the sun had already commenced changing it into a slimy mud. As we ground to a stop, our 26 x 4 tires were chewing into the stuff and there we were. I was

33

provoked and I said to "Red," 'What in the name of common sense did you land here for?' He rather meekly replied, 'Gosh, I didn't know it would be like this—I just thought this would be a good place to clean out the carburetor.' He then explained that he thought the motor was slightly rough and he could quickly drop the pan, unscrew the jet, and remove whatever might be restricting the flow of fuel.

"We carried lots of tools on this type of jaunt and "Red" was an excellent mechanic. Sure enough, when he pulled the jet, there was a little cluster of fuzz stuck to it.

"In the meantime, I had been sliding around in the mud attempting to figure out how we could leave the place. The little plateau was quite small in area, with deep inclines dropping from all sides. As I stood on the edge peering in every direction, there wasn't a sign of life nearby. Not even a curl of smoke was perceptible.

"I went back to the plane, which by then was all buttoned up, and I asked "Red" just how he expected to get out. Obviously perplexed and chagrined, he of course had no idea; but he said, 'If we could find a little lumber, we might be able to build a ramp and get rolling again.'

"Having previously noted the remains of what appeared to be an old cabin, we decided to go down and have a closer look. By this time we were covered with mud but we still wore the overcoats and derbies. We slid down the slope to the pile of debris which probably had been there for years. Scratching and pawing into what once had been lumber, we finally found a few pieces that looked more solid than the mud. We dragged them up and onto the plateau. It was very discouraging.

"We then started the motor and attempted to get the plane closer to one side of the plateau, hoping we could turn it around and possibly take off in the opposite direction. We tried to use some of the pieces of wood as pry-bars but they broke almost as quickly as we attempted to use them. Then "Red," who was a big fellow and strong, said, "You get in and try to rock it loose while I lift on the struts.'

"During this operation, with the tail banging up and down, it finally broke loose and the plane started to roll. Turning my head quickly, I expected to see "Red" hanging onto one of the struts but he was just standing there, wildly waving his arms for me to keep going. I pulled the throttle and stopped. "Red" came running and asked why I hadn't kept going. I countered that I couldn't pull out and leave him. He argued that he would have found a way out somehow.

"By this time the plane was fairly close to the edge and as we stood there looking over and down, "Red" said, 'You know, if we could get this thing rolling and over the side, I believe we could get going fast enough to get into the air.' For a moment I was silent. Then I said, 'And what will you do when you hit that watercut down there?' The latter was perhaps 70 to 80 feet from the top and about two feet wide. "Red" said, 'We'll jump it—I hope!" Then as we both continued to peer on down and beyond the watercut, I said, 'And what are you going to do if we're not off when we get to those trees?' The latter were a heavy ring at the botton of the slope. "Red" could only project that 'If we're not off by then, we'll slam into them!'

"This was food for additional consideration. We discussed our chances for survival and the insurance on the plane. I explained that we were covered for crash but the plane had to burn in the crash to collect. We discussed our baggage. Should we remove it if we could, following the crash, or would it look better to let it remain inside. We decided on the latter.

"Memories are made of this!" says pioneer pilot Karl Voelter of Coral Gables, Fla., as he ponders that long-ago mountain-top landing by "Red" Jackson!

During his famous endurance flight with Forest O'Brien in 1929, Dale (Red) Jackson went out on this precarious "cat walk" daily under all conditions to inspect and tune up their Curtiss Robin's Challenger engine. This dramatic close up shot was made in the air on the 15th day of flight.

"Then to get better traction, we let the air out of the tires and once again we cranked up the engine. I had already decided that "Red" would be the pilot. It was his idea. Again we pushed and shoved while revving the engine and finally we were just at the edge. I climbed in back, holding a package of matches in my hand and got a good grip on the assist straps. We were both in a sweat.

"The noise was terrific as we rolled over the side. Rocks and stones were crunching and flying and the airplane sounded like it would disintegrate. We gained speed and, sure enough, as we came to the watercut, "Red" bounced the plane across! Our speed increased and so did the noise. Then "Red" pulled back hard and I can still hear the tail-skag as it smashed into the rocks. We were airbourne—not solidly, but at least we were holding our own. Quickly we levelled off and with a tremendous shout of joy, we had it made! We flew on to Syracuse where we knew Gordon Hood, our manager there, would be waiting.

"With no air in our tires we would have to make a short approach; but slips to a landing were not uncommon in those days. As we came to the field, we noted a tremendous crowd. This was unusual on a Sunday morning. And it was even more unusual and certainly surprising when after we had landed and stopped, we noted people handing each other what looked like paper money. Then they saw us and we noted their surprised looks. Some were handing back the money. When Gordon Hood came up beside us, he quickly explained that only an hour before, one of our pilots, "Red" Devereaux, had taken off through an overcast in a Robin exactly like ours, and the same color. Devereaux was attempting an altitude record for OX's. Many on the field had bet he would be back within the hour. And then we came in. There were two aboard our plane. Devereaux had gone off solo. (Note: Devereaux set an altitude record for OX-5's that day which, as I recall, was 16,700 feet.)

"After replacing our tires and having lunch while mechanics checked and fueled our plane, we were off again for Akron. It was my turn to fly. "Red" was in the rear thinking up tricks to play, as was his usual custom. He would hold the door open with one foot. Then he would reach forward and shut off the fuel and let the motor conk out and holler, 'Turn on the gas!' (The valves were on the floor on each side of the cabin). Finally, he picked up the bag of tools and threw it on the hardwood floor. I thought sure we had blown apart.

"We made it to Akron uneventfully and the next morning after a wash and polish, we demonstrated the airplane to Goodyear officials. They bought it, demanding that same airplane. That was the start of Goodyear's vast air fleet. This first airplane would become a test plane for their new doughnut aircraft tires. And with that same plane that we had rolled over the side of a mountain, they soon set a record of 608 (bounce) landings in one day.

" "Red" and I went back to New York that night by train—with "Red" trying to figure out a trick to play on the engineer!"

THE SEAT OF THE TROUBLE!

Whimsey again enters the picture in the following experience of one of this nation's better known airmail pilots. In his own words: "That much-quoted adage, 'a man pays for his pleasure,' has ever held true whether from over-enjoyment of dated bottles, viewing go-go gals from the third table front, or from merely bringing 'em back alive from the unexplored jungles of Sumatra!"

Thus, seeking a momentary respite after hours of dull flying was responsible for the biggest boner in the flying career of E. Hamilton (Ham) Lee, veteran mail and passenger pilot, and might have easily been the cause of that worthy name finding its way into the columns entitled, "Obituary Notices."

Lee, who a few years back hung up his cap after 36 years of hauling mail and passengers for the U.S. Air Mail Service and later the United Air Lines, comes across with an experience herewith he felt best to wait this long to tell. And since a story is usually best told by the person involved, here is "Ham" Lee's personal description of what he dubs his prize fiasco:

"I pulled my prize fiasco many years ago, way back in February of '22 while flying the mail from Chicago to Minneapolis. As usual, I landed at La Crosse to have the Standard E-4 I was piloting serviced, then took off on the last leg of a wearisome grind in the cold open cockpit of my Government mail plane.

"It was bitterly cold, about 23 degrees below zero; and while a strong Northwest headwind cut down my speed considerably—still the weather was nothing to fret about. I flew low in order to avoid as much wind as possible and to make the going a trifle less rough.

"It was crystal clear, the objects below standing out sharply like pictures propped at the end of a stereoscope. Even scenic beauty can become quite monotonous, so when I spied a dredge in the distance, frozen fast in the ice of the Mississippi, I hailed it with keen interest and enthusiasm.

"What perverse trick of Fate prompted me to pick that particular moment to seek a bit of amusement, I can only explain by my extreme boredom.

"I suddenly dived lower until I was about level with the dredge roof. The men aboard who, but a second before had gazed in hypnotized wonder at this marvel of the ages, came out of their stupor and began waving their arms wildly before scuttling to the safety of the cabin.

"My laughter was short lived as the explanation of their fear and the telephone wires hit me simultaneously. I immediately pulled the ship up between two and three hundred feet, then cast a quick apprehensive glance over the side. The impact of the wires had sheared away the two struts supporting the outboard wing which began to show an affectionate interest in the ground below. A peculiar sound from the front of the plane drew my attention away from the collapsing wing. The propeller was merely voicing its disapproval of entangling telephone wires by imitating the weird noises of a peanut vending machine.

One of the best known airmail pilots on the transcontinental rte. in post World War I days was E. Hamilton (Ham) Lee, shown here in his official U.S. Gov't Air Mail uniform. It was during this period that Lee did his "dredge buzzing". He retired many years later as a United Air Lines captain.

"Although there was very little choice in landing spots on those rugged flats of the upper Mississippi Valley, something told me to bring her down while I still had even a slight say in the matter.

"I brought her gently to an even keel with the ground and congratulated myself a moment too soon. If the trees, which loomed up ahead, had been spaced three feet further apart, I would have made it. I left the wings cluttering up the landscape and came to an abrupt halt as a low hanging limb forced its way through the fuselage. The impact threw me forward off the seat; while the rough branch ripped across the side, greedily taking with it the rear section of my pants, not to mention a goodly portion of my epidermis and all of my dignity.

"I had already pulled the bags out of the mail compartment and was still swearing loudly at the results of my folly when I discovered that I was not alone. A bearded guardian angel, smoking a corn cob pipe, and driving a span of sway-backed mares hitched to a bob sled, had witnessed the mishap and hurried to the rescue.

"In a very short time my mangled pants were replaced by a pair of

39

bluejeans and I was standing at a kitchen sink scoffing off numerous cups of black coffee.

" 'You ain't the first pilot what's landed 'round these parts,' my host informed me. 'It warn't more than six months ago one landed right out yonder.' He indicated the location by nodding towards a small flat field in the distance. 'Only,' he continued, 'you come down a mite heavier. That pilot stayed a short spell on account of the weather and larned me a lot about flyin'.'

" 'How soon can I get a train North?' I interrupted.

" 'One leaves Winona in about an hour. You want I should drive you over?' asked my rescuer.

" 'I'd sure appreciate it,' I said dismally, my feelings still at a new low. 'Golly, I've sure made a mess of things.'

" 'Yep,' said my host, eying my backside knowingly, 'yer rump's in pretty bad shape.'

" 'If that was all I had to stew about,' I retorted, 'I'd be sitting pretty. It's explaining the crack-up that has me worried.'

" 'It warn't yer fault, son,' said the old man earnestly. 'I seen the whole thing and I'd swear it warn't yer fault.'

" 'What do you mean, not my fault,' I asked skeptically.

" 'Wall,' drawled my guardian angel, 'that other flyer larned me a lot. He told me right confidential that you boys all fly by the seat of your pants, so how in blazes can anyone blame you when you didn't have no pants on your seat and damn little hide!'

"As the mail and I headed for Minneapolis on the express, I chuckled to myself. I was half tempted to spring the old boy's alibi on my boss but common sense warned me to hunt around for an excuse which sounded, at least, a little more plausible. After all, it was a good job and I did enjoy eating."

HOT PANTS AT THE LANGMACK HOMESTEAD!

David Francis Langmack, a Lebanon, Oregon, pilot more commonly known as just Dave, was quite a believer in an old postal service slogan that he had conveniently altered to suit his own purpose:

" — — — no snow, nor rain, nor day, nor night shall stay this pilot in his flight."

In fact, on some of Portland's blustriest, rainiest days with the clouds settled down over the hills like a giant gray curtain, days on which even the birds were walking, the bark of an OX5 Jenny would be heard overhead and such Portland pilots of the day as Tex Rankin, Frank Anderline and John Langdon would merely look at each other, without a glance from the hangar, and say:

"Here comes Langmack!" And sure enough here came Langmack!

On occasions when Dave would finish hauling passengers at country fairs, chautauquas or what have you, it was invariably pitch dark—and, it being Oregon, just as invariably raining. He'd merely look the situation over and the decision was generally "go," which meant head back to the Langmack farm at Lebanon. Then he would hand his brother, Charlie, a flashlight so he could read the gasoline gauge and keep him posted on how well the fuel was holding out.

He wasn't at all worried about landing on the family farm after dark and in the rain. He had a simple procedure on which his entire family was

Probably the most storm-swept Jenny in the (then) forty-eight states, and it's intrepid pilot, Dave Langmack, of Lebonon, Oregon.

well-briefed. Three or four burlap bags, a jug of coal oil and a box of matches were standard equipment in the Langmack woodshed which adjoined the house. When Dave's ship was heard approaching the farm after dark, one or more of the family would grab the burlap bags, the coal oil and the matches, run out into the pasture and light a fire at a prescribed spot. Dave would then come in and land adjacent to his merrily dancing "runway light."

On one of these occasions Dave arrived over the family farm as unexpectedly as the second coming of the Lord. It was one of those coal black nights one would associate more readily with the meanderings of the ghostly hounds of Baskerville rather than with an airplane. The overcast was exceptionally heavy and the rain coming down in torrents. The entire Langmack clan was sound asleep when out of the Stygian darkness Dave whams over the house in his OX5 Jenny at just about midnight!

Now this midnight buzz job had the comparable effect of a huge charge of dynamite going off in the town square at 3:00 a.m. The whole Langmack family hit the ceiling at once!

In spite of their sleep-befuddled minds, the one thing they could collect in their thoughts was "get the bonfire started!" But, instead of the old burlap bags that were laid out for that purpose, Dave's mother grabbed the first thing she could get hold of in the closet. Meanwhile his dad grabbed the jug of coal oil and the matches, missing the burlap sacks entirely. By the time Dave and his Curtiss Jenny had completed his circle of the farm, the Langmack "lighting system" was working in good order and Dave landed without further ado.

A Sunday or so later, however, Dave decided it might be wise to attend church with his parents to keep his standing with the Lord in good stead. Going to the closet for his one and only good suit, he found only the coat and only then made the sad discovery that, instead of burlap bags, his mother had grabbed and started that midnight bonfire with his best Sunday go-to-meetin' pants!

All of which only goes to show that a pilot doesn't always lose his shirt—sometimes he loses his pants right along with it!

42

A HELLUVA WAY TO RUN AN AIRLINE!

Airmail pilots, the derring-do glamour lads of the 1920's and 30's, came in for their fair share of humor and horseplay.

Back in 1926 airmail pilots of the Pacific Air Transport, one of the parent companies of the now giant United Air Lines, often would be given extra duty after their regular mail runs, such as flying doctors to emergencies at towns not readily accessible by road, or hopping vaudeville actors to their next one-nighter and so forth. And thus, one typical bay area day—weather-wise—it became the lot of pilot Johnny Guglielmetti, a regular on the main run from San Francisco to Los Angeles, to haul a public relations man from San Francisco to Santa Cruz, a distance of 70 miles or so.

With fog creeping over the hill at San Francisco and fog already reported at Santa Cruz, Guglielmetti took off in his mail plane—a Ryan M-1 open-cockpit monoplane powered with a 200-horsepower Wright J-4 air-cooled engine—with his not overly enthused passenger cramped into the front cockpit.

Since parachutes were generally worn by airmail pilots, Guglielmetti had strapped his 'chute on the public relations man. The catch was, this was the only parachute at hand so Guglielmetti went without.

As they approached the pass through the Santa Cruz mountains at Los Gatos, Guglielmetti could see the fog pouring over the top of the hills from the Santa Cruz, or ocean, side. Question: Should he pull up over the fog or try to duck through the pass under it—both unhealthy any way one looked at it. At that instant the ticklish decision was made for him by the engine quitting completely dead. Not even a sound left!

Looking for the best spot to make a forced landing, Guglielmetti spotted a side-hill wheat field, fully planted, and stalled his ship down perfectly. He immediately jumped out of the cockpit and took a look at the engine but could see nothing amiss. He then put his hand on the prop. Nary a speck of compression! He could turn the prop with one fingertip. The timing gear had sheared off!

About that time his passenger said, "Is this Santa Cruz?"

"Hell, no!" replied Guglielmetti.

"Well, what are we doing here?"

"The damned motor quit!"

"The motor quit? The MOTOR quit?" sputtered the aghast public relations man. "What kind of a flying service IS this?"

Guglielmetti just gave him a disgusted look and stalked off to the nearest farm house where he placed a call to his Superintendent of Operations, Grover Tyler.

"Where are you?" asked Tyler. "In Santa Cruz?"

"Hell, no!" replied Guglielmetti. "I'm down in the mountains about half way between Los Gatos and Santa Cruz!"

"Is your passenger all right?" was Tyler's next query.

"No—I jumped out!"

"JUMPED OUT?" shouted Tyler, almost crawling through the phone.

"Well HE didn't have a parachute," answered Guglielmetti—and he could hear the horrible gurgling gasp at the other end of the line as Tyler braced himself against the wall to keep from fainting.

Of course Guglielmetti, feeling a sweeping surge of sudden remorse for his practical "joke," hastened to explain that he had already landed and stopped when he jumped out—and, anyway, that it was he, Guglielmetti, who was 'chuteless and not his fuming, sputtering passenger who still thought it was one helluva way to run a flying service.

"Here's Johnnie now!!" —former Pacific coast and transcontinental airmail pilot, Johnnie Guglielmetti, as he looks today. He is superimposed in front of a photo of the original Ryan M-1 monoplane he used to fly on the coast mail run in 1926, and the ship that figured in the forced landing described in this story. Guglielmetti is also a retired TWA airline captain.

NO CONSIDERATION WHATEVER!

Practically coincidental with the preceding incident, and along about the same time, southern California pilot Bob Lloyd went against his own better judgment and consented to take a passenger over the Tehachapi ridge from Los Angeles to Bakersfield in what even then was considered an obsolete "crate."

Bob just barely cleared the ridge; in fact, he almost rolled his wheels on top. Just as he was thanking his lucky stars for getting over on the down side, his engine quit deader than the proverbial mackerel. He glided with the slope of the mountain to a tiny ravine into which he slid, knocking off the wings and all other external fittings and protuberances. But he managed to bring the wreckage to a standstill amid piles of rock and shale at the base of a cliff without a scratch to either himself or his passenger. The passenger sat very silent for a moment, gloomily contemplating the situation, then turned to Lloyd and blurted:

"You might at least have glided down the *other* side of the mountain where I could have caught a bus!"

ONLY INVERTED, TAIL-FIRST
DEAD-STICK NIGHT LANDING IN HISTORY!

Making an inverted tail-first landing, with the top of the rudder acting as a tail skid, is not an easy maneuver in the first place. In the second place such an arrival on terra firma with little or no damage to the airplane is something slightly less than miraculous and probably a stunt that would have been more than welcomed by such acrobatic aces of the era as "Speed" Holman, Tex Rankin, Freddie Lund, Gordon Mackey or Johnny Livingston!

Nevertheless such a landing was successfully, though quite inadvertently, made by Jenny-thru-jet pilot Charles J. Langmack (Major, USAF, ret.) of Lebanon, Oregon, in 1927. Moreover, it was accomplished after dark!

Langmack had purchased a wrecked Waco-9 biplane from a friend who apparently had made a good landing with it against the side of a barn. He spent the better part of a year rebuilding it, knocking off periodically in order to get a job to enable him to purchase more supplies and parts.

After completing the project, Langmack (as most aviators did in those early days) embarked on a barnstorming tour of the state, hauling passengers on 5- and 10-minute hops for $2.50 or $5.00 per ride.

On the night the tour ended, Langmack flew passengers until after dark, it being against any pilot's religion to turn down money so long as it came in. About nine o'clock the purses apparently ran dry and Langmack took off for the 100-mile hop back to Lebanon. It was a beautiful night—full moon, not a ripple in the air or a cloud in the sky—a great night for flying so he just

The only known pilot to make a perfect three-point tail first landing upside down by moonlight - Charlie Langmack, of Lebanon, Oregon. Later a major in the Air Force.

relaxed, listening to the OX-5 engine purr and watching the exhaust flames dancing in perfect unison from the short stacks.

There was a small field just north of town that he had used many times over a period of years. Arriving over Lebanon he circled the field prior to setting down. There was enough moonlight for him to easily determine the field's boundaries, so he chopped the throttle and headed in on his final approach. Everything was going fine and he cleared the boundary fence all nicely slowed down with the stick well back for a perfect three-point landing. The next thing he knew there was a terrific impact out of nowhere and the engine quit at the same time! Everything was deathly silent for a few seconds. The lights in the surrounding homes and buildings had disappeared. He knew there would be a second impact for he was still in the air. But the second impact never came. The Waco-9, now upside down and going tail first, sailed ahead another 80 feet and gently settled to the ground so softly that not even a hole was torn in the cloth covering of the upper wings!

Skidding to a stop, there Langmack hung upside down in his safety belt not knowing for sure what was straight up or straight down. At this point he did what most pilots do in such situations that they shouldn't—he flipped open the safety belt, whereupon gravity immediately took over, resulting in a good-sized knot on his noggin!

Checking the ship over after regaining his equilibrium, Langmack found the only damage was one bent propeller tip, a bent rudder top—and, of course, the knot on his head.

Further investigation showed that, in his absence, some local "sportsmen" had built a trap-shooting platform smack-dab in his touch-down area, and had driven two heavy posts about three feet high to hold it in place. This he had hooked with his landing gear as he cleared the fence and was about to land, causing the ship to flip upside down and continue tail first to its "perfect" landing!

THIS FLIGHT WAS A CORKER!

While Walt Bohrer, co-author of this compilation of aeronutical histerics, has never had the dubious pleasure of flying in combat, he has experienced being in an aircraft being riddled by "flak"—from the inside! How could that happen? Well, it wasn't easy! Here's the story right from the "horse's mouth," as one might phrase it—unless, of course, another word is used to substitute for "mouth!"

"Flying a Wright 'Whirlwind'-powered Waco 10 biplane, we were on what we dubbed a 'brainstorming' tour of southeastern Oregon's high desert country, the 'we' being yours truly and Hans Mirow who later became one of Alaska's better-known bush pilots.

"Both of us were dangerously curious by nature. We later totally wiped out the Waco by flying it through a juniper fence while watching a group of Diamond-O Ranch cowboys rounding up some wild horses. That was in the Donner und Blitzen Valley.

"This time we were flying between Malheur and Harney Lakes and what was getting the best of our curiosity was some unusual activity going on in a hay camp below. What it looked like from 1,500 feet was an open-air booze party. A celebration of some sort was taking place and since it looked like there was plenty of room for two more participants, we landed.

"What we found was a celebration all right but not one that was prompted by a bumper crop of alfalfa hay. The ritual, rather, was in honor of a bumper crop of top quality home brew! And since this was in the prohibition era (1929), we immediately offered our unlimited assistance.

"The natural wind-up was that we took all participants on their first airplane ride in return for which we received one full case of a dozen quart bottles of home brew packed in wet sawdust!

"Being anxious to get on up to Vale, a desert town some 150 miles northeast, we took off with our 'prize' and yours truly in the open front cockpit.

"The late afternoon was hot and it wasn't long before we hit what is now termed turbulence, but which in those days we knew only as 'bumpy air.' We were being bounced around like a cork in a fast running creek when all of a sudden, 'WHAM!'—a sharp explosion followed immediately by another! Something hit my leg and I knew we had collided with a brace of wild geese! However two more immediate 'WHAMS' told us, 'Duck! The beer is blowing!' Corks and glass were ricocheting around that front cockpit like flak. Home brew was being sprayed about as if from a fire extinguisher. I was not only soaked with it but plastered with wet sawdust, and to make matters worse, my leg had been cut by jagged, fast-moving bottle fragments! With the continuing explosions of our precious cargo, my only safety lay in hooking my heels over the edge of each side of the cockpit. In the meantime German-born Hans had cut the throttle and, in his excitement, was yelling in German: 'vas ist los?' (What's the matter?)

"Well we finally arrived at Vale after a most uncomfortable hour and a half. Ten bottles of brew had blown. What a loss! And to add insult to injury, the remaining two bottles were grabbed by 'Bake' Russell and Howard Maish who had just set down ahead of us at Vale in a Ford tri-motor!

"As Tex Rankin later put it: 'Aw, you fellows just flew through an 'alcoholic front!' "

Pilot Hans Mirow, parachutist Pat Paxton and co-author Walt Bohrer looking for elusive passengers on a Pendleton, Oregon wheat field in 1931. Mirow later became a noted bush pilot.

BASIL ROWE AND THE
"SUGAR CANE PATROL!"

Almost everyone has heard of the famous "dawn patrols" of World War I, but not so many have head of the "Sugar Cane Patrol" of the mid-1920's and therein lies a tale.

More often than not, barnstorming pilots of the period were forced to fall back upon sheer ingenuity in order to keep gasoline in their tanks and biscuits heading in the general direction of their stomachs. Not that most pilots were starving, you understand—that came later during the depression; it was just that they dearly loved to eat!

In the latter part of that roaring decade ugly rumors, rife with dire predictions, were emanating from our national capitol. The newly formed aviation division of the Department of Commerce—so went this gossip—was about to pounce upon all barnstorming and air show activities and fold them up like a campstool.

Naturally, this took in most of the seat-of-the-pantsers then flying as they were all barnstorming, performing stunt-flying or air-racing. So many of them began casting about for means of flying livelihood securely beyond the clutches of this new and imminent woe. In order to obtain this goal, a few of the more ambitious pilots even left the country!

One of these enterprising fellows was Basil Rowe who, at this writing, is still revving up in good shape in Florida. Off to the West Indies went he—out of the Department of Commerce's clutches all right, but into—well, let's just let Rowe relate his own story:

"The year 1926 was a hectic one for the aviation fraternity because of the threatening legislation that was anticipated from Washington and the many rumors that supposedly originated from the newly created C.A.A. There would be no more barnstorming, all operations had to be conducted from permanent bases; there would be no more exhibition and stunt-flying, and so on went the sad news. I had become fed up with all the scuttlebutt so I had decided to sell all my racing and barnstorming ships and head for the West Indies as soon as the races were over at the Philadelphia Sesquicentennial Exposition. At that early stage the United States had no aviation laws for the control and regulation of flying so the Federation Aeronautique Internationale of France was accepted as the controlling authority for the air meets. In order to compete in the races all pilots were required to obtain a license from this organization. That year Orville Wright was Chairman of the Contest Committee in the States and had an office in Philadelphia where I went to renew my license. My number was 223 and when he handed me the license he said that before the year was out he hoped to have 300 pilots. Today the FAA has 700,000 licensed pilots!

"I had very good luck at the Sesquicentennial meet and sold the SVA, Martinsyde and Thomas-Morse at a good price, largely due to the good performance of the ships during the meet. I also sold the two barnstorming Avros and bought two new Wacos and a complete new barnstorming outfit

Basil Rowe's Sugar Cane Patrol "fleet" at Santiago de los Caballeros, San Domingo in the West Indies in 1926. The fleet consisted of two 100 H.P. Waco-9 biplanes. The "gasolina" came free for the advertising.

and shipped the entire mess to Porto Rico. I knew it would be some time before the newly created CAA could give me any trouble in that far-away hideout and I could keep ahead of them by changing islands or countries if and when they caught up with me.

"It didn't take long to work over the little island of Porto Rico and next we moved to Santo Domingo for daily exhibition flights at the International Exposition at Santiago de los Caballeros. After the exposition closed, we moved to Santo Domingo City where I made a contract with the foreman of the Barahona Sugar Central to patrol their cane fields. He dispatched his chauffeur to the Central with orders to have a field ready for us the following afternoon. During the harvest season large numbers of Haitian field-workers were imported to assist in cutting the cane since the local labor supply was insufficient. They, of course, brought along their Voodoo which was far more essential to them than their machetes. Since the season was limited, the cutters frequently took advantage of such a splendid opportunity to hack more wages out of the rich Americans. If the owners were reluctant to comply, the cutters strengthened their arguments by setting fire to the ripe cane. The Central was then obliged to accede to their demands or lose that portion of the crop. Unless it was cut immediately, it was ruined.

"When we dropped out of the sky onto the newly prepared field the next day, news had already got around that the airplanes would arrive. The whole town had turned out for the event. I had made a contract with the Texaco Company to advertise their products throughout the West Indies. They had painted the two Wacos with their house colors and insignia until they looked like flying gas pumps. The foreman informed the workers why the Texaco Birds were there. He said he meant business. I was glad nobody asked me what he meant because I had no idea of what we were going to do beyond buzzing the fields. But while the foreman stood in the seat of the Waco telling of the many horrible things that would happen to the field-workers if they gave any further trouble, painting me as the Blackbeard of the air, Bill Wade, the other pilot, was getting ideas.

"He started on it that night. The first step was to drink a lot of beer. Soon he had a sufficient stock of large, empty liter bottles. Into each he stuck a piece of dynamite or poured some gunpowder with a fuse attached. He leered blearily when the last one was done and yelled, 'bombs away!'

"When we went out on patrol next day, he ducked down into the cockpit and touched his burning cigarette to the frayed end of the fuse and tossed the charge overboard. It had barely passed below the bottom of the ship when it exploded with a shock that nearly blew Bill and the ship right out of the air! He let all the bombs go. There was no more trouble with the cane cutters—at least no more trouble between them and our employers. The cane cutters forgot about higher wages and began to concentrate on Bill Wade and me.

"We were quartered in one of the Central bachelor houses, complete with cook, during our patrol activities. One evening when I crawled under the mosquito net and pulled back the sheet of my bed, I was startled to find a doll spattered with blood. It had been crudely whittled from wood and grotesquely painted to make it appear as hideous as possible. Obviously it was intended to resemble a person and before I'd finished looking at it, it began to look like me! Attached to its shoulders were the crumpled and broken wings of a bird which gave it the appearance of some kind of winged monster. The body was pierced with thorns and nails. The natives believe

A much later photo of our "Sugar Cane Patrolman" - Capt. Basil L. Rowe, for years the top Pan-Am pilot.

52

that with those ouangas (charms) the houngans (priests) can produce the most painful form of death to a victim without any personal contact whatever. Of course the victim is notified, as I had been, that he is marked for death.

"The first thing that I had observed about the doll was that it was headless. I didn't have to be told the significance of the warning. The worst thing conceivable to a Haitian was the possibility that his body would lose its head, without which it would forever be excluded from Voodoo heaven.

"As I gingerly picked up the doll, I heard a far-away throbbing of Voodoo drums, as though my action had set them in motion. Their boomety-boom-boom, boomety-boom-boom chiseled its way along my spine to raise a lot of prickly splinters. I had heard tom-toms before, but they had just been an interesting form of primitive musical instrument. Tonight, however, they seemed to carry a threatening message that scraped my nerve ends. They had a hypnotic, feverish rhythm unlike any other drum beat. I felt a little shiver work itself through my frame as I heaved the ouanga through the open window.

" 'What was that?' asked Bill, holding a shoe that he had just removed from one foot, poised in mid-air.

" 'Just a little obie,' I said as I crawled under the sheet, drawing the mosquito net tightly together. Secretly I wished the net were lined with a good grade of armor steel since the windows were as open as the blue yonder.

" 'Maybe I better sleep with my boots on,' Bill said, starting to replace the shoe. 'I suppose we will have to listen to that the rest of the night.'

"The next evening we found a trussed-up white chicken hanging by its wings from a string attached to the electric light fixture in the ceiling. The shadow from the struggling fowl crept across the floor like the shadow of death.

" 'What's the idea?' Bill said, studying the bird closely.

" 'I don't know.'

"Bill started to cut the bird down and carried it out into the kitchen. The discovery started the drums again.

"The third night our black magic took a more gruesome turn. Before crawling under the mosquito netting, I carefully looked over the bed. There was a lump under the sheet. It wasn't entirely a surprise. What sort of grim object would it be this time? I knew it was some live creature for I had seen a slight movement. I very carefully eased back the cover, fully expecting a snake to pop up its head. A baby iguana blinked its eyes at me in fear. His head darted from side to side while he clawed at the sheet with his front feet in an attempt to run away. It was a futile effort because his hind legs were wrapped up over his back and the long toenails inserted though a slit in the skin just above the spine.

"I decided I'd had enough black magic. I was beginning to worry about our ships parked out in the open field. The next day I went to see the foreman and told him the story.

" 'So that's the reason they're pounding the drums!' He looked worried as he reached for the telephone and said, 'Bring those potions here. I'll send

53

for the papaloi.' (Papaloi is the local term of Haitian creation meaning Big Shot, Top Dawg or Supreme Authority upon whom all the natives live in awe and respect because of his command of witchcraft. Whatever he wishes is a command, murder or what have you.)

"'Well—uh——that will be rather difficult. Last night we ate the iguana and the night before we ate the chicken.'"

FLYING—ALASKAN STYLE!

It's a sure bet that a pilot can't barrel around such rugged country as Alaska very long without running into some interesting experiences. Bush pilots will tell you a lot of them but you can also be sure there are a few, too, they'd hate like sin to let out.

For instance, a pilot landed one day in a Waco seaplane with a bunged-up float. When he got out, he promptly announced that while taking off, a whale had come up to "blow" and he'd hit it! A mechanic, taking a closer squint at the damage, saw wood splinters wedged in the hold and yelled back at the pilot:

"Hey! That whale you hit—it had a LIMB on it!"

What's a "bush pilot?" We're glad you asked!

An Alaskan bush pilot is one of those rugged guys who accelerated transportation in the land of the midnight sun by replacing dog teams and sledges with aircraft.

He's a rough 'n' tough flyin' bozo who loves to get up at all hours of the night, or to lie down and sleep any place he happens to wind up the following night. He loves coffee. Has to, because coffee is the first thing offered wherever he lands. He eats everything from "mucktuck"—that's dried fish, Eskimo style—to seal liver. And he can take just about anything the North country can dish out in the way of weather.

The rank, if it can be so called, of bush pilot is not bestowed on him by honorary degree; nor is it earned by flying in the northern bush any set length of time. Rather, the title, or reputation, of bush pilot is earned only by long years of bitter experience and hard work.

Alaska is a strange place at best. It's a hard drinking, hard working and hard playing country. A trapper can come into town and go on a glorious binge for a month, but let him see a pilot drunk and he'll charter another plane every time. Alaska is proud of its pilots. You'll often read that they are the world's best. Alaska's proof? "They're still flying, aren't they?" or, "They've got to be—look at those mountains!"

And the places they fly to! A map of Poland looks mild in comparison. Try these for size, then, try to imagine a train or plane announcer calling them without his esophagus getting tangled with his east tonsil: Unalakleet, Shismaref, Mumtrakmut, Allakakat, Kuskokwim, Napakiakmute, Talkeetna, or Kulukak. Yet every one of those names are as common to the bush lads as Chicago, New York or Atlanta to the airline boys. And if you want to toss in an extreme, there is Eek!

Just to prove the point that anything can happen to a bush pilot, take the case of Alaskan pilot Archie Fergusen flying through freezing rain en route to Kotzebue in a Fairchild monoplane with two passengers. The ship started icing up and as the ice coating grew heavier, the plane began losing altitude.

Describing that not uneventful flight later, Archie said, "I told those passengers, 'You know, if we keep losing altitude like this, we'll hit the next ridge and crash.'—and you know what? We DID!"

Then there was the sad case of bush pilot Oscar Winchell who dropped what he thought was a load of meat to a snowbound family up Candle Creek way only to discover later that it was a phonograph and 300 records! All but two of the records were broken and the people played those over and over for months!

RUSS MORGAN OR FREDDIE MARTIN?

It has been said by some that each flight made by a bush pilot is either an adventure or near catastrophe—or both. Even a supposedly blasé, routine flight can sometimes develop into something utterly ridiculous—as in the case of the following "routine" flight made by bush pilot Lloyd Jarman:

"I was dead-heading (riding free of charge) to Seattle in a DC-3 that belonged to a non-scheduled airline operating out of Seattle. The airplane was in good shape but some of the pilots were just borderline for Alaskan operations. Some good and a lot of bad. However, even with a few near misses the airline had a fair flying record. I had gotten stuck in Anchorage because of a foul-up of air crews with the company I worked for. One of our planes had a bad engine and the rotating of the pilots and mechanics for trips to "the outside" was thrown out of sequence, so I had bummed a ride with another outfit.

"We took off from Anchorage about 2:00 p.m. heading for Cordova to pick up a load of canned crab and salmon for Seattle. I thought the field in

There are times when one bush pilot would like to punch another bush pilot in the schnoz – and Lloyd Jarman now a Seattleite, is still thinking about it.

Cordova was a little short for the load that was put on the airplane but I figured the captain knew his airplane. On take-off we picked up speed as the tail came up but the ship felt real sluggish and the end of the strip was coming at us awfully fast. Just beyond the end of the runway was a good size stand of alder trees. I was standing in the aisle and had a better look at things than the two sitting pilots. We lurched into the air and here came the alders! The captain signaled 'wheels up' and the co-pilot complied. We picked up a little more speed but the alders were still higher than the airplane—and looking bigger every second. Both pilots were disgustingly relaxed and poker-faced when we went through the trees, the props chewing treetops up into small pieces and spitting them in all directions, and the branches banging back over the plane—and we kept right on going! The captain adjusted the prop controls and neither said a word. I finally figured out that this was not unusual for them. For me, yes.

"We were fairly light on gas so a stop was scheduled at Annette Island to refuel. It was getting dark so I pulled out an old mattress and laid on top of the freight to catch a little sleep. There is nothing as boring as riding behind someone else on a routine flight. I must have slept at least three hours when I was suddenly awakened by the props speeding up. Then I heard ice hitting the fuselage as the props threw it off. The air was very rough and the crab and I were bouncing around as though we were riding on a buckboard over a rough mountain road, neither the crab nor I being tied down. I stood up and looked out the front cabin window. The weather was so thick that the wing tip light did not show even in the plexiglass reflectors installed for that purpose. 'WHAM!' Ice hit the fuselage again. This time it was a little louder and the props had run up a little longer. I made my way up to the cockpit. The windshield and side windows were a mass of ice. Every light in the cockpit was on and it was as bright as day. The co-pilot was fast asleep, slumped over in his seat. The pilot was really working. We were off automatic-pilot as the air was too rough. He was a real picture of 'How to Wear Yourself Out In Two Easy Lessons!' Banking slightly left, then right, he would then reach up and crank the radio installed in the overhead structure. With one hand he would press the earphones to his head. Then, like a desperate man, he would reach over and dial some more of the radio gear under his window. Again he rolled the control wheel. Again he pressed his earphones to his head with a baffled look on his face. I could feel the blood draining down away from my head. The only thought that came to my mind was, 'This guy is lost! Somewhere along the line, while goofing off, he lost the beam!' Our gas stop at Annette was critical as we were light on gas out of Cordova due to the overload of crab, fish and so forth. I had visions of us ditching somewhere along the wild, desolate Alaskan coast as the engines quit from lack of fuel. Looking back the captain finally noticed me standing in the cockpit doorway. Like a beaten, desperate man he pushed the headphones away from his ears and looked up. His lips moved but it took me a couple of minutes to comprehend his words.

" 'I know damn well that's a Los Angeles station but for the life of me I can't tell if it's the Cocoanut Grove or the Paladium!'

"Right then I could have killed a pilot!"

THE OLD LAMPLIGHTER!

Nearly all of us at one time or another have been "assisted" by a "mother's little helper"—the kindly soul who means well but more than often just gets in the way, thus slowing down progress, or worse yet, manages to get everything nicely fouled up!

Such was the case experienced in 1954 by U.S. Game Management Agent Ray Tremblay of the U.S. Fish & Wildlife Service, Bureau of Sport Fisheries & Wildlife, while on duty in Alaska. Since the assistance rendered almost cost him his eye teeth, Tremblay has forgotten nary a detail of the episode and relates the following as proof:

"In 1954, during a moose season patrol, I was operating a Gull-Winged Stinson off the strip at Paxson Lake, Alaska, which, at that time, was only about 1,500 feet long and crowned by a hill at the north end. This significant obstacle prompted most landings being made to the north and take-offs to the south.

"Two fellows had to be flown to Healy one evening which meant a late return flight. I knew I would be arriving back after dark and the only possible landing would have to be towards that hill and had better be precise and on the very first part of the strip. Just to add a bit more zest to this fun-type adventure, the landing light on the Stinson was not working.

"Never trust a guy with a lantern!" says U. S. Bureau of Sports Fisheries and Wildlife pilot Ray Tremblay of Alaska, shown here with his gull-winged Stinson. And this story tells you why!

"Before departing, I requested one of the temporary workers to put a gas lantern on the exact end of the usable runway on the left-hand corner, which would act as a guide for my descent and touch-down point.

"As anticipated, it was quite black when I returned; and, after several passes, the lantern appeared and finally became stationary at the end of the field.

"I made a long approach, my eyes on that lantern; came up even with it; chopped the power, and hauled back on the stick ready for touch-down. Suddenly a sickening, sinking sensation surged through me as the bottom fell out. The airplane stalled and plunged for what felt like a considerable distance and an eternity passed before hitting the ground. Needless to say, what was supposed to be a feather-light landing turned into a bone-jarring, soul-quaking crash which would have destroyed any other gear than the Gull Wing's. After gingerly bringing the plane to a halt and finding out how many teeth I had left, if any, I shakily walked back to find out what had happened.

"Nothing much, EXCEPT for the fact that my young friend had REALLY wanted to help me out. To do so, he had attached the gas lantern—my handsome guiding light—to a 20-foot pole which he then proudly held high into the air above his head!"

"AH, SO! NOW TO HITCH-HIKING!"

Flying hunters, Japanese style, will never be bush pilot Norm Helwig's forte!

For the past several years Helwig has been flying out-of-state hunters from Anchorage, Alaska, to Susitna Lodge, 160 miles northeast—a nine-hour drive by car but a mere hour-and-ten "hop, skip and yump" in Helwig's little Cessna 180.

One late August day in 1967, a non-pilot guide from Fairbanks phoned him and said he had a couple of Japanese hunters arriving at Anchorage from Tokyo the next day and he wanted Helwig to fly the three of them up to Susitna Lodge.

This was fine with Helwig. The more business the better, but he informed the non-pilot guide they would have to take off for the lodge no later than 1:00 o'clock in the afternoon for the simple reason that, at this time of the year, the days were rapidly shortening and the weather in the mountainous areas deteriorating at about the same pace. He also admonished the guide to have his Oriental charges travel as lightly as possible as he, too, might have supplies to fly to the lodge, as he very often did. From this point we turn the story over to Helwig:

"Sure enough, the next day the guide called again—but not until 1:00 o'clock, the pre-set take-off time limit. His Japs were in, he said, but they were still uptown and couldn't possibly make it out to the airport until 3:00 o'clock. At 3:00 p.m. he called back and said the trip would have to be postponed until the following day—his Japs were on a buying spree that wouldn't quit and he couldn't get them out of the stores! Again I admonished him they had to keep the load down.

"At 5:30 the afternoon of the next day they finally showed up in a rented station wagon at my tie-down area at the International Airport, and it was truly a sight to behold! That station wagon was literally bulging with guns, cots, stoves, lanterns, sleeping bags, hibachis and every other conceivable piece of outdoor equipment—plus a case or two of sake! This, in itself, was funnier than hell when you have a Cessna 180 for transportation.

"After introductions, which were very congenial and polite, I explained to them there was NO way to get all that crap in my little 180. This entailed another loss of precious daylight while the guide explained to the one hunter, who spoke only broken English, who in turn explained to the other hunter, who spoke no English at all, that we could take only the bare essentials.

"After about 15 minutes of what closely resembled a Keystone Cops slapstick comedy, I got the two hunters strapped in the rear seat and proceeded to stack four new rifles with scopes between them across the front to rear seats. Then with the baggage compartment stuffed full, the rest of the stuff was stacked around their legs and feet and behind the rear seat. The lodge was out of bread so the coup de grace was a case of bread stacked on their laps! This, too, was very funny as, from the outside of the airplane, only two Japanese heads could be seen sticking above all that gear! And, of course, they couldn't see each other because of the rifles.

"After another 15 minutes' delay while the station wagon was returned to the rental people, we finally limped out to the take-off ramp. (For the benefit of the F.A.A., we never overload our airplane; we were just at gross.) Being an excellent bush plane with plenty of power, the good old 180 lumbered down the runway and staggered into the air.

"Normally the flight is a direct route from Anchorage north to the east side of Talkeetna, over the mountains and through Stephan Lake Valley. Along through here we usually see a number of black bear, caribou, moose, mountain Dahl sheep, and occasionally a grizzly.

"As we approached the lower end of Stephan pass, the sky was rapidly darkening and the black clouds and fog hung down into the trees. This eliminated the direct route.

"Our flight so far contained a rather fluid flow of running conversation between the guide and myself, and even an occasional Oriental utterance from the obscured passengers.

"Our next route was to fly west to the Anchorage-Fairbanks route of the Alaskan Railroad, then north to the Susitna River pass. Naturally it was getting darker by the minute and the weather rapidly was getting worse. For the most part, this portion of the Susitna River runs through a deep canyon about 100 yards wide with banks at least 300 feet high. There is one hairy, narrow, sharp turn about three miles from the railroad.

"It now had begun to drizzle, further cutting down visibility, and fog-type clouds were hanging at the top of the canyon. As we started up the river, the conversation lessened perceptibly. As a matter of fact, it ceased! We finally broke out in the upper end of Stephan Valley and it looked much better. Again the conversation sputtered as we neared the valley of the Susitna Lodge.

"As we rounded the bend heading for the lodge, a great black rain cloud appeared in the darkening twilight directly over the lodge landing strip. The conversation again abruptly ceased. It was pouring down rain and completely dark!

"I buzzed the lodge and they wasted no time bringing out two cars, one on each end of the strip with head lights on. Due to the extreme turbulence from the wind blowing down two valleys and converging directly over the small strip, plus the gross load in the airplane, it was necessary to make a high speed approach. The first pass was so hot, it was impossible to land. After two more passes, I figured this had to be the time. Just as we were about to touch down, a sudden gust blew us clear off the runway! I could hear the brush hitting the wheels. In a split second I could visualize 180 Cessna, bread, Japs, gear, guide and me scattered all over the landscape from hell to breakfast! After applying full power and the Lord's Prayer, we became airborne! This was a subtle clue to abandon the idea of landing in the disastrous wind and rain. During this season of the year there are many road hunters driving on the adjacent Denali Highway. This helps immensely for night flying down the road toward the Summit Airport, about 50 miles to the west.

"I never have seen anything so beautiful as the 3,000-foot, lighted Summit airfield! During the last 45 minutes nary a word was uttered, either English, Japanese, or profane! After landing and taxiing to the tie-down area, the engine purred to a stop and the Japanese flowed like wine. The hunter sitting directly behind me, who spoke broken English, patted me on the shoulder and said:

" 'Ah, so! You very fine pirot! Yamaguchi say tomorrow we go by automobile!' "

ANY OLD SPUD IN A STORM!

With 24 hours of daylight during the summer season in Alaska, a bush pilot might be in the air at any hour as business dictates—and no bush pilot is about to lose any business.

Thus it was that pilot Bob Byers happened to be flying along the Yukon River in the wee—albeit bright—pre-dawn hours when his engine suddenly konked out. Out of gas!

What had happened was that his airplane was of a type equipped with rubber gas tanks and with that type of plane, the gas all sucks out if the gas cap comes off, and that happens every so often if the cap is not twisted on properly. In this case pilots revert to the old adage that necessity is the mother of invention and use a potato as a gas cap until a replacement is found. Well, as you may have guessed, Byers lost his gas cap—and his gas—and there he was down on the Yukon in the middle of nowhere with nary a potato in sight!

Keeping his ship off shore away from snags and other obstructions, Byers floated down the Yukon until he came to the village of Tanana.

Despite the early hour, an Eskimo lad was down at the river bank. Byers tossed him a rope and the boy pulled the ship into shore.

"Hey, boy!" yelled Byers. "I need a potato for a gas cap. See if you can find one some place!"

The Eskimo lad takes off in high gear to locate a spud in the village while Byers commenced refueling operations with the five-gallon cans of gas he carried in the ship.

Meanwhile the native boy knocked on the door of the first cabin he came to.

"Who's there? What do you want?"

"Byers needs a potato!" was his answer.

"Go away—you're drunk!"

Next cabin, same procedure: "Go away, you are drunk!"

After five cabins the boy finally got his potato and Byers was able to take off, all of which goes to show that even the lowly spud can save aviation in some areas!

AIR-DALE OR SKY-TERRIER?

The "doggondest" thing to take place in many a moon was the day an Eskimo gal's dog went to pooch heaven as an unannounced passenger on a pontoon of bush pilot Dean Goodwin's seaplane.

Goodwin, operator of the Goodwin Air Service at Juneau, had just put in at a little native village of Hoonah for an accumulated load of Uncle Sam's mail for the "outside." Being the dead of winter when visiting aircraft in such places are as rare as false walrus teeth, every Eskimo and his dog—and we mean that literally—came down to the float for a look-see.

With a snowstorm threatening to cut him off, Goodwin quickly tossed in the mail and gunned his ship away from the float. At that precise moment, and unseen by Goodwin, a dog broke away from the crowd and leaped on one of the plane's pontoons. Hanging on as if his very life depended on it—and, brother, it did, Fido rode out the icy take-off and managed somehow to stick to his precarious perch as the ship was airborne. Goodwin's first glimpse of his unscheduled passenger was when he circled back over the wildly waving group of well-greased natives 500 feet below. At the same time his canine customer spotted home and made a flying leap into space—and dog heaven.

There's a little Eskimo miss up there in Alaska to whom pilot Goodwin has promised a new dog but one with anchors attached!

Veteran bush pilot Dean Goodwin of the Goodwin Air Service, Juneau, Alaska, thinking about lunch on the Taku River.

"TAXI!! — OR ANYTHING!"

Passengers unwittingly provide no end of entertainment. For instance, on one of pilot Dean Goodwin's trips to Ketchikan, several salesmen, handling competitive lines of merchandise, were aboard. One salesman in particular was fidgeting around like a sinner at revival meeting. Obviously he was devising ways and means of beating his rivals into town. Now as this particular incident occured during World War II when pilots were firmly cautioned not to use their radios for non-essential business, Goodwin's reply to the "expected" note from the salesman to radio ahead for a taxi was that the only thing he could possibly radio ahead for was an ambulance in case of an emergency. However, he added, were the salesman to become suddenly ill, he could have transportation awaiting him upon arrival.

The result was that he had never in all his flying career seen a man get sick faster than that drummer boy! The guy should have been in Hollywood! He groaned and moaned and writhed the entire distance to Ketchikan. His competitors were so worried that they helped him up the gangway, even going so far as to carry his luggage!

Sure enough, awaiting him was the requested ambulance with two white-coated attendants in charge. And to the complete chagrin of the supposedly-dying salesman, there also stood not one, but THREE taxicabs! If the salesman had set a record becoming ill, he set a new record recuperating! In fact, he became so angry, he refused ALL transportation and WALKED into town!

EAGLES AND AIRPLANES DON'T MIX!

Then there was that beautifully clear day when pilot Goodwin, flying along minding his own business, caught the flash of something out of the corner of his eye. In the same instant there was the damndest crash he'd ever laid ears to. The whole airplane, one of those old Winnie Mae-type Lockheed Vegas, shuddered and shook like a nude Ubangi on an ice floe! A UFO? Well, hardly—Ken Arnold hadn't spotted those first ones yet.

A shower of plywood fanned by a terrific blast of air gyrated about the cabin. Coming up off the floor of the ship, Goodwin slowly opened his eyes—one at a time, fully expecting to see a wing gone. Both wings were still there. "Goodwin, old pal," he said to himself, "you got no tail!" And he had just convinced himself of that sad fact when he happened to glance up at that section of the cabin immediately adjacent to the Goodwin noggin. There, where the radio transmitter should have been, was a huge jagged hole through which showed a goodly sized hunk of Alaska! And well padded about the edges of said opening were the feathers and whatever else it takes to make up a late bald eagle! Had the huge bird elected to commit suicide twelve inches further to the left, Goodwin's skull would have been branded with a permanent imprint of the emblem of America's invincibility, and some poor CAA inspector probably still would be probing through the wreckage for the fan dancer who lost all those feathers!

Bush pilot Goodwin shows where the near-sighted eagle smacked his Lockheed Vega monoplane.

CANINE CLOWN OF THE NORTH!

From one of the preceding narratives of the north country and the following saga, it would seem all Alaskan pooches are suicide prone.

Pilot Elbert E. "Al" Parmenter of La Crescenta, California, a former Alaskan bush pilot, was based at Copper Center, Alaska, during the winter of 1939 while flying for the Cordova Air Service. Together with his wife, Ruth, and baby daughter, Jean—the only white child in many miles at that time, Parmenter lived in a small house near the airport. There he would take off several times a week with the necessities of life for many people in remote places. Each evening or so he would return with passengers, orders for groceries, gold pokes to be delivered to the Copper Center Bank and ad infinetum, which means just about anything else one could think of. Of this he recalls that "I had to be very careful with the jugs that were in the grocery orders—I never broke a one!"

One crackling cold day Parmenter landed his Bellanca Pacemaker monoplane at McCarthy, site of the once giant Kennecott Copper Mine.

Former Alaskan bush pilot Elber E.(Al) Parmenter
at Valdez, Alaska in 1938. Note dug out skis
sitting on boards so they will start easier ----
and Arctic Circles under Parmenter's eyes!

There a big game hunter and trapper met him and asked if he could be flown to one of his camps in the Nabesna country east of the Wrangell Mountains. As it developed, there were several complications to this deal and they were all wrapped up in one name—"Whitey!" He was to fly the trapper to his camp all right, which was well enough except for the fact that he was to take his dog team back to Copper Center and care for it until he could pick up the trapper again several days hence.

This team was comprised of several big good-natured black and white Huskies with nary a care in the world. They all knew what hard work was on the trail but they were well cared for and happy as clams at high tide with their lot—especially their leader, the big clown, "Whitey."

When Parmenter landed back at Copper Center with his lively cargo, his wife, Ruth, came out to help him get them settled in their new temporary quarters—the airport hangar. He gave each of the Huskies a dried salmon for supper and saw to it that they all had a drink of water before it froze.

Ruth and the dogs became great pals at once, particularly after she fed them with table left-overs, and an occasional juicy bone which would disappear in one great gulp.

One morning while Parmenter was away on a two-day trip, Ruth was busy preparing a moose roast for dinner and baking some sourdough bread. She suddenly had the strange feeling that she was being watched. She turned toward the window behind her and there was "Whitey," his feet on the window sill and that big "dog smile" on his face. What to do now, she thought. He had broken out of the airport hangar and she was afraid he might get lost.

She put on a coat and raced out the back door, picking up a dog chain en route. "Whitey" came tearing around the house, jumped up and gleefully licked her face as she put the chain on his neck. He then took off for the hangar at high speed with Ruth hanging onto the chain. The footing not being so hot on packed ice and snow, Ruth's feet went flying out from under her and she was practically pulled like a sled for a half block, or the equivalent thereof, but she finally succeeded in anchoring the grinning lummox back in the hangar. She was uninjured and, of course, had a ball telling Parmenter of her wild experience upon his return to Copper Center.

The day finally arrived when Parmenter was to return his canine charges to their owner. The passenger seats were removed and the animals tied in their places. The cabin door was hinged across the lower end so it could be opened over the strut and would stay there for loading. He took off and circled back over the house at 200 feet, tipping his wings as a farewell to Ruth who was waving from the yard. "Whitey," looking out the window spied her and made one tremendous lunge at the door, poking his head and half his body through the flexible bottom! Had it not been for all of the combined strength of one of Parmenter's arms and the chain, old "Whitey," too, would have dived right on out to that great dog kennel in the sky! After the first bounce, that is!

AND THEN THERE WAS "BOOSTER!"

Yes, and then there was "Booster," a wing-walking, stunt-loving fox terrier "partner" of Charles A. Lindbergh back in the general's barnstorming days.

This was during those pre-Paris hop years when Lindbergh was stunt man and mechanic for "Shorty" Lynch, one of his first instructors at Lincoln, Nebraska.

Booster joined the Lindbergh-Lynch barnstorming team somewhere in Kansas. Lynch gave Booster his first hop. In order to keep the dog in the cockpit of their ship—a "Standard," a rope was fastened around the pooch's neck and tied to the seat. All went well and Booster seemed "airminded" enough until Lynch started his glide for a landing. When about 50 feet above the ground, Booster stuck his nose out and spied a rabbit. That did it! Out he bailed—for the full three feet of rope! Lindbergh nearly had a strangled dog when the ship finally taxied to a stop. Artificial respiration brought him around and thus ended his first lesson.

On more than one night hotel desk clerks scowled with suspicion at a somewhat bashful young man who apparently was smuggling something up to his room under his leather coat. Once inside the room, Booster would slide out from under his master's jacket and stretch happily on the foot of Lindy's bed.

After a few weeks' air-time, ol' Booster became so "plane-broke" that a stout harness was rigged up for him. It had a few inches of leather leash on the belly band and a snap on the end was secured to a ring on top of the Standard's turtle-deck. And thus, atop that turtle-deck outside and behind the rear cockpit rode Booster for many an air mile.

Booster's lessons progressed apace with his learning. At length he was given his first loop while riding the turtle-deck. He lived through it and learned how to brace his legs just right against the fuselage, taking the slack out of the leash and riding through the loops without the slightest whimper. And he progressed until he "soloed" on a tailspin. From then on life was easy. Booster finally graduated to wing-walking, slowly edging out to the last strut and back to the cockpit.

Booster became "prop wise" early in his career. When Lynch or Lindbergh would go up without him, he'd wait until the ship stopped taxiing. Then he'd warily skirt the nose of the ship, watching the propeller out of the corner of one eye, dash around to the tail, hop up on the horizontal stabilizer and trot on up the fuselage to the turtle-deck. There he would vainly pose for photos with a very smug look on his face.

The "partnership" finally broke up in Bird City, Kansas, when Lindbergh told a group of boys that he'd trade Booster for another dog. In five minutes 50 boys had returned with at least one dog apiece—either their own or someone's else.

Who can tell but what Booster, in later years, trotted up to a picture of Lindy while taking a stroll with a girl friend and said, "Lindbergh? Sure I know him—used to barnstorm with the guy!"

GRAHAM "GROUNDS" HIS FOGGIA FARMAN!

No book of this nature would—or should—be complete without at least one aerotale of World War I—the truly romantic era of aviation. The romance lasted as long as weaponless aviators of both sides of the fracas saluted each other as they flew past, but when one sneaky pilot could stand it no longer and bopped his rival on the "dumkopf" with a brick as he flew over, the romance ground to a screeching halt. After that came pistols and then the rat-a-tat-tatters through the props, and all hell broke loose from there on out.

But every now and then, both in combat and out, something amusing would happen to break the tension on the various tarmacs.

As an outstanding example, there was the sad case of "Red" Graham, the pole vaulter of Chicago U, the bird that cracked the only good ship on the line at Camp Sud in Foggia, Italy.

Quite a passel of our American lads were trained at Foggia in 1917. Red, a carrot-topped, sawed-off, tight-knit stripling of 22 was one of them. Slower than Louisiana molasses, lazy as a government mule, a complete teetotaler and world's pole vaulting champ, all Red wanted to do was fly. He had grinned himself through ground school; was constitutionally opposed to drill, calisthenics and all other forms of physical exertion, and had slept all the way to Italy. The sub alarm off the coast of Ireland disturbed not one whit his bunk fatigue on the poop-deck of the S.S. Mongolia, nor did the beauties of England, France or Italy raise a flicker of his reclining eyelashes. But Red did fly naturally, soloing in less than three hours and doing his eights, one-eighties and cross-country without scratching a tail skid.

One bright mid-April p.m. Red went for his altitude test. The old blue-nosed "flying bathtubs" the Italianos furnished at Foggia for the altitude check were the cabin jobs of the day—old Farmans with a 90-horse Anzini engine mounted directly behind the pilot who was perched out front in a round-nosed nacelle that swallowed a short guy—like, say Red.

Well, the barograph was set going, sealed and tied in the front cockpit. Red put on his tin britches, his leather flying jacket and the ornamental crock of cork, leather, steel braces and what-have-you that made up a helmet to protect his dome and off he took!

Now the old Farman could climb three or four hundred meters in a couple of hours—if you lifted hard enough on the stick. She had gas for about two hours and used castor oil for lubrication that left an odor for miles behind which made poison gas smell like attar of roses.

Red was strong for records, being the pole-vault champ, so he nursed the Farman to the last inch of its ceiling and held it there until the final drop of fuel sneezed through the wheezing Anzini—castor oil smell and all. With a dead stick he headed back for the field, two and a half miles down and ten miles west. There wasn't a breath of air. With that height he could have made a field 15 miles away. The field was two miles square and as level as a frozen mill pond.

Down came Red in one easy sweep—a perfect approach. On the east side of the field was a gravel pit and adjacent to the gravel pit, a heavy string of telephone lines, a typical airport even today!

Without varying a gnat's bristle Graham glided straight into the wires, nose first. The ship hesitated just an instant, as if deciding whether to proceed or stay perched. And then the wires broke! The Farman did a half floperoo and piled up, right smack in the bottom of the gravel pit!

The ambulance roared, the siren screamed, and everyone rushed for the wreck. Out of the dust and wreckage crawled Red, a few tattered bits of fabric and spruce wrapped around his neck, the seat of his tin pants completely torn out, the crock helmet cockeyed over his left ear. All in all he was one sad looking bird. Not a busted bone or a drop of blood, but an exceedingly dumb, bewildered expression on his decidedly peaked map.

"What the hell happened, Red? Didn't you know those wires were there? Didn't you see that gravel pit? What the hell?"

"Well, you know," said Red with a sheepish grin, "I was just flying along there thinking about my gal, when blooey—she went! The next thing I knew, I had the center section as a collar and sand running outa my ears!"

WATCH OUT — OR TIME FLIES!

Back in the good old daze of aviation during the early thirties, Freddie Lund was not only chief test pilot of the Waco Aircraft Company of Troy, Ohio, but he was also a top air show attraction the length and breadth of the country. Freddie was equally adept at air racing, both cross-country and closed-course, i. e.: racing around pylons, but his real forte was acrobatic flying. He was, for instance, the first civilian pilot to perform the outside loop, a maneuver in which the pilot "rides" the outside of the circle, a massive "G" force on both aircraft and pilot. The feat had previously been performed only in military aviation by Lieutenant James H. (Jimmy) Doolittle of the then Army Air Corps.

Backed by this acrobatic prowess and his general flying skill, Freddie's performances were ever being sought by air show promoters throughout the nation.

It was to attend one of these events that Freddie, accompanied by his pretty blonde wife, Betty, an accomplished pilot in her own right, was

The flying Lunds relaxing between performances at the 1929 National Air Races, Cleveland, Ohio.

75

cruising along 2,500-feet over the northern Iowa countryside in his taper-wing Waco biplane. This happened to be one of those warm, lazy midwest PM's that would make any flight over flat farmland disgustingly ho-humdrum and, although within spitting distance of his destination, Freddie decided to remedy the situation with a few snap rolls.

Glancing at one of his two wristwatches to check the time (he always wore two watches for time comparison), Freddie proceeded to start his snaps. Now Lund was justly proud of one of his wristwatches since it was an award for winning an acrobatic event a few weeks previously. It was a Benrus with his name and the description of the event for which it was won inscribed on the back.

Somehow, during his series of monotony-breaking snap rolls, the Benrus flew off his wrist and disappeared over the checkerboard farms more than two-thousand feet below. Freddie almost bailed out to retrieve his pride and joy, and may have done so had it not been for Betty in the control-less front cockpit.

Sick over his loss, Lund proceeded half-heartedly to the air show wondering why he had done such a stupid thing. He tried unsuccessfully to console himself with the thought that it could have happened any time. But, true to "show biz" tradition that "the show must go on," Freddie, despite his loss, turned in one of the best performances of his career.

Upon landing, a grizzled old Iowa farmer broke through the huge air show crowd and, spitting a stream of high speed tobacco juice at a passing sow bug, asked:

"You Freddie Lund?"

Freddie Lund in front of his famous tapered-wing Waco.

Of course it mattered not to the farmer that the name "Freddie Lund" was emblazoned across the entire underside of the upper wing, or that, during his performance, his name was constantly being mentioned over the field loud-speaking system.

"That's right," answered Freddie, "I'm Lund."

"Then this is yours!" said the farmer, and to Freddie's utter consternation, handed him the prized Benrus wristwatch he had lost from the air some twenty miles away. And—even more amazing—the Benrus was STILL running!

"Where in blazes did you find this?" asked the elated but thoroughly stunned Lund.

"The dang thing dern near hit me on top of the head while I was out doing my plowing," said the farmer. "I picked it up and saw your name on the back and recollected seeing in the paper that you wuz going to pull some airplane stunts over here. So I jus' drove over with it—thought mebbe you'd like it back!"

Freddie likely would have kissed the guy except for the fact that he let drive another stream of high speed tobacco juice at a second passing sow bug. The least he could do, since the farmer refused a monetary reward, was to offer him a ride over his farm, which he happily accepted.

"Next time," said Freddie, "I'll watch out and keep my watch in!"

ANYONE FOR SWAHILI?

It's generally conceded that language is no barrier but it sure hampered the style of several pilots we've heard about. Bob Ceniceros, for instance. Bob was one of the pilots to whom nothing ever happened until World War II. By December 7, 1941, he'd proudly logged 2,000 hours. This was enough time to include him in the exclusive clan of Tex Rankin's pilots teaching Air Corps cadets at Tulare, California.

From the Rankin Aeronautical Academy, he went on to bigger things—the Air Transport Command—and to Ceniceros, this was really big! Up to now he was a wide-eyed native-born Californian who'd never seen the snow or even a large river with water in it and an assignment of traveling all over the world delivering airplanes was a bit overwhelming to say the least. He literally licked his chops over this prospect but his anticipation extended beyond the cold, white stuff and large bodies of water flowing hither and yon. He was thinking of all the girls he'd meet! He could speak Spanish like a veteran as Ceniceros wasn't exactly an Irish name, so he was sure he had it made. He'd be able to communicate and make points with the lovelies in all of the foreign ports.

His first schedule took him to Morocco where his mission was no sooner completed than he straightened his tie and embarked on his quest for dark-eyed beauties. To his utter dismay, he found the local lassies spoke only French and Egyptian—nary a word of Espanol!

His next mission took him to Algeria where he was sure he'd have better luck. And what did he find? The girlies spoke only Arabic!

His next hop took him to India where he carefully made the rounds only to find he should have learned any of a score of dialects of the Indian language. There was nary a Spanish-speaking gal in all of India! In fact, during his whole tour of duty, he never met a soul who could speak Spanish! He couldn't even understand the French or Italians although their languages contained Spanish words.

It was a most unhappy situation for this Spanish fly boy but finally the great day came. Ceniceros was assigned on C-54s to the Casa Blanca-South American run. After a long year in Africa, he figured at long last he really had it made. He eagerly looked forward to fresh fruits, fresh milk, and most of all, a bevy of Spanish-speaking beauties, with him in the role of Errol Flynn.

Upon landing at Natal, Brazil, he streaked for town and struck up a conversation with the first chick he saw—and guess what!!! The one damn country in all of South America where they spoke only Portuguese!

From the above it is no surprise that Ceniceros got a "D" in high school geography!

Like many of the American pilots arriving in la belle France during the first World War, Tex Rankin was one who got into immediate trouble through his total ignorance of the French language. Impressed with the beauty of the mademoiselles in the Romarantin area where he was stationed, Rankin mentioned his feelings to his barracks' "buddies."

Former international aerobatic champion, Tex Rankin, with a few of his many trophies including the International Aerobatic Trophy, front row center. Airplane is a Ryan S-T.

"Don't tell us how good-looking the gals here are," they told him. "Go tell the gals! They love compliments—especially from American soldats!"

"But how can I?" asked Tex. "Other than 'Merci, ma'm'selle,' I don't know one lousy word of French!"

"We'll teach you what to say!" chorused several of his barracks' mates. "What choice bit do you wish to convey?"

"I want to say, 'You are very pretty, mademoiselle. Will you go out with me?' " replied Rankin.

"Oh, that's a cinch," they said. "All you have to say is 'Bon jour, ma'm'selle! Voulez vous couchet avec mois?' (Good morning, mademoiselle. Will you go to bed with me?) Now practice that a few times and try it out on one of the French nurses on the post!"

Tex did just that and having it down pat, boldly strode up to one of the nurses and proudly demonstrated his newly-acquired knowledge of the French language.

The nurse gazed at him in wide-eyed amazement for one short moment, and then grabbed the nearest broom and, in a tyrade of unquestionably pure French, started beating the astounded Rankin over the head with it, accompanied by the howls of glee of his so-called buddies!

Needless to say, Rankin consulted other more reliable sources for further linguistic training!

Jimmy Doolittle, too, was a pilot who encountered the language barrier.

Forced down in an Alpine valley while flying a Curtiss Hawk across Switzerland on a stormy day, Doolittle landed in a soggy cow pasture and, crouched beneath the wing, waited the storm out.

The rain finally stopped and the ceiling lifted enough to permit a take-off but a heavy farm wagon was smack-dab in the path of his take-off.

Doolittle tried to budge the wagon with no luck. Looking about for help, he spied a tramp with a bindle stick watching him from a fence next to a road.

In turn, Jim tried his broken German, Spanish and French in an attempt to get the fact across to the knight of the road that he wanted the wagon moved, but to no avail.

In sheer desperation, Jim led the tramp over to the wagon, put his hands on the shaft and indicated forward motion with his head.

"Oh, hell!" said the tramp. "You want me to help you move this wagon!"

Doolittle, in his linguistic desperation, had completely forgotten to use his native tongue!

The tramp turned out to be a wandering citizen of Spokane, Washington, and they spent another hour hunched under the wing of the Hawk discussing baseball and politics back home in the States.

GENTRY FINDS HER PLACE!

Shortly after she had qualified for membership in the Royal Order of Flying Jackasses in the spring of 1929, aviatrix Viola Gentry found herself in line for an "oak leaf cluster" to same, if such were given!

We hasten to explain that the Royal Order of Flying Jackasses was founded in 1926 by one L.B. Rawlings and, to our knowledge, the Order still stands. Mr. Rawlings, 'tis said, was a New York pilot who, while testing airmail pickup devices at Curtiss Field, saw so many foolish things being done by experienced pilots, as well as fledglings, that he founded the famous Order—or DISorder! The prime membership qualification was to pull a whopper of a boner far beyond the call of inanity!

Miss Gentry's chance to make a blazing entry into the Order came early in 1929 while she was load-testing a new plane preparatory to an attempt to establish a new endurance flight record for women. She had established such a record the year before, but as it had since been bettered by other women, she was anxious to set a lasting record.

Her ship had 60 gallons of gasoline in a 100-gallon tank installed in the front cockpit, while the wing tanks were full. The extra tank had been loaned to her by (now General) Jimmy Doolittle.

Having been cautioned by practically everyone and his dog on the field to shut off the gas in event of a forced landing, this advice was, naturally, first on her mind when her motor quit dead on take-off when she was but 15 feet in the air. She shut off her gas all right but—as General Doolittle so aptly phrases it—ran completely out of experience! Not only did she fail to remember that she was not heading into the wind but she also forgot to correct for drift. Result: a smashed propeller and a wiped-out landing gear, plus a public presentation of a stickpin set with a small gold jackass—her membership into the Royal Order of Flying Jackasses!

Barely a few weeks later Miss Gentry was engaged to deliver a new Arrow Sport biplane to the East Boston Airport for Hugh Morton, a Boston attorney who had purchased the craft.

Now while all this sounds simple enough, such was not the case. Since air maps were unavailable, road maps had to be used and pilots would always try to follow a road, river or railroad track to make sure they were on course.

At Mitchell Field, Miss Gentry was told the prevailing foul weather was slated to clear up later in the day. The clouds finally commenced breaking up about 1:30 p.m. and Miss Gentry, accompanied by a well-known New York pilot of the day, Bill Ulrich, took off for the city of beans and tea parties. Their intentions were to head out over Long Island Sound on a direct route to Boston. However good these intentions were, they were foiled by further foul weather in the vicinity of Riverhead and were forced to return to Curtiss Field.

After an hour's wait a second start was made—with the same result. They were turned back at Greenport by clouds so low they would have had

Viola Gentry with her endurance flight partner, Bill Ulbrich (right), and her mechanic, Carl Schneider. Pictured in New York in 1928. (N.Y. Daily News photo)

to cross the Sound at less than 300 feet, a situation over which neither she nor Ulrich were overly enthused.

Back again at Curtiss for refueling, she was given gas, another weather check and a few "asides," the undercurrent of which seemed to be: "Well, what can you expect with a girl at the controls?"

That did it!

The fact that the birds might be walking that day wasn't about to deter THIS gal from going to Boston! At least she would get the heck out of Curtiss even if they had to land at Yonkers!

After consulting with Ulrich, she decided to fly overland to Worcester and there spend the night with friends, going on to Boston the next day. Thanks to Frank Hawks, the Worcester Airport sported a brand new beacon, the first in the East, and this was her big chance to try it out.

Off again for the third time, Miss Gentry started to climb above the overcast just after passing Oyster Bay. Should trouble develop she could always sit down at Madison or New Haven—ANY place but back at Curtiss Field! Near the Connecticut shore a light drizzle started to fall, indicating an early darkness. She had been driving Ulrich batty with a constant barrage of questions as to whether he was sure they were on course and Ulrich kept assuring her they were just that. In spite of all this, Miss Gentry came down to a 1,000 feet so they could more closely check the map points and be sure they were headed for Worcester. Finally she was positive they were near and commenced circling, certain she would sight the new beacon any minute.

The gas gauge was in another nose dive by this time. In fact, it was within a fraction of empty so she decided finding a likely looking landing spot might be the better part of valor. Just as the gauge indicated absolute empty, she spotted a large farm field near the road and headed into the wind for a landing.

Well, they lit—with a crunch wholly atop a large mound, the salient features of which were a sudden braking effect and a horrible odor.

To see an airplane land in those days was quite an event in most places and this seemed no exception. A crowd had soon gathered, mostly children and a few parents who had run across from some houses and large buildings on the opposite side of the field. A few cars had also stopped along the roadside across a high fence.

Knowing they would have to be picked up, Miss Gentry decided to call Lieutenant Irwin McWilliams, manager of the Worcester Airport, if a phone could be located. One of the parents from across the field said her phone was available, provided Miss Gentry possessed the wherewithall to cover the cost of calling.

Leaving Ulrich to guard the tilting aircraft atop the smelly mound, she accompanied the woman and a battery of youngsters to one of the houses across the field. There she was met by a host of curious visitors—and six policemen!

After an indeterminable number of rings, the airport finally answered. Lieutenant McWilliams, they said, was at the hospital where his wife had presented him with his first baby earlier in the day. A call to the hospital was made but McWilliams had already left. However they supplied a couple of phone numbers, one of which was the Worcester Telegram. The Aviation Editor there advised Miss Gentry to sit tight; he would locate McWilliams and call back.

While awaiting this call, Miss Gentry decided to call the New York airplane agency that had hired her to deliver the plane and let them know what had happened. The secretary in New York asked where she was and Miss Gentry asked one of the policemen. She nearly dropped dead when she was told she was on the grounds of the State Insane Asylum, five miles from the Worcester Airport! Furthermore, she and Ulrich would have to be identified by some reliable person before they could leave the grounds!

It was after 8:00 p.m. before McWilliams called. He had just returned to the hospital and received her message. He'd be right out to identify and pick them up as soon as he could locate his lawyer since some damage may have been done to the grounds or the airplane.

At long last they were able to leave. By this time they were practically starved to death—and by this time, too, Miss Gentry had been told by Ulrich that they would need a new propeller, and that the vile smelling pile they had landed on was a large heap of ripe stable manure!

"At last," replied Miss Gentry, "I am where I belong—on a pile of manure in an insane asylum!"

HOW ABOUT THE CROWS?

The foregoing story brings to mind another instance involving a State Hospital in the same general area, an airplane pilot, a forced landing, and the Aviation Editor of a New England newspaper. We present same to wit:

Indians generally are known for their longevity, but it was up to pilot Mel Hodgdon of Middletown, Ohio, to discover it was aviation—or the lack of it—that caused the red skins to live to such a ripe old age!

In June 1919, while a pilot for the old Whittemore-Hamm Company,

The mustachio'd aviator is dashing young Mel Hodgdon, circa 1919, then pilot for the Whittemore-Hamm Co. of Boston.

Hodgdon was assigned to fly his friend, Ted Hedlund, Aviation Editor of the Boston Post, from Saugus, Massachusetts, to a small aviation meet on Cape Cod. As they crossed over the city of Boston at about 3,000 feet, the radiator of their plane started to steam and blow hot water back into their faces. There was naught to do but search for a place to return to Mother Earth, and do it fast! Fortunately, they spotted a nice green field within easy gliding distance and had no trouble making a smooth landing.

On the ground, their first act was to lift the engine hood and determine the cause of the trouble. The bottom of the radiator had cracked and allowed most of the water to escape. The engine was a four-cylinder, 90-horsepower Hall-Scott which was famous—or infamous—for its heavy vibration, which eventually had cracked the radiator.

As they were determining what course of action to take to seal the crack, a man walked toward them from a large brick building at the end of the field. As he approached, his first words were:

"The Indians never had anything like this and they lived to a ripe old age!"

Hodgdon and Hedlund agreed with his logic, though a bit startled by his strange sense of humor.

The stranger then proceeded to give them a filibusterous discourse on the high state of civilization of the American Indian. As he rambled on, he seemed to become more and more incensed that anyone would attempt to fly while the Indians did not. Finally he grabbed the wing struts and began shaking them furiously while continuing his tyrade on the lack of aviation throughout Indian history. As Hedlund and Hodgdon attempted to restrain him, another man hurried out from the large building. He spoke quietly to their "humorist" and pointed toward the brick building. The Indian lover then turned and trudged off.

After explaining their difficulty to the new man, they asked him where they had landed.

"At the Boston State Hospital for the Insane," he replied, "and that was one of our inmates, but he is harmless."

Needless to say, Hodgdon and Hedlund did not get to the aviation meet at Cape Cod that afternoon, but they did learn that the early Indians never had forced landings!

CERTAINLY NOT FLY-LIKUM-EAGLE!

And while on the subject of redskins, we hasten to point out that not all Indians are capable of outwitting aviators, such as was the case in the preceding episode. Sometimes it is the other way around.

Colonel Edward V. Pettis, USAF (ret.), former Department of Commerce aviation inspector in the Pacific Northwest, loves to tell about the incident of Tex Rankin and the big Indian at Bend, Oregon.

As always, Tex was appearing at an air show at which Pettis was the covering federal aviation inspector. And as was very often the case during the 1930's, upon the completion of his aerial antics, Tex would haul a few passengers at a cent a pound. For this purpose he would carry a small bathroom scales with him.

On this particular day a big Indian, apparently from the nearby Warm Springs Reservation, showed up on the scene and asked to be "weighed in." Upon noting the needle of the scales disappearing from sight and deciding the Indian weighed "about a hundred pounds less than a horse," which on this hot afternoon at Bend's fairly high elevation could prove disastrous, Tex told the Indian he was sorry but he had to quit flying for the day in order to get into town for a reception at the Bend Elks Club—which, naturally, there wasn't. But the Warm Springs warrior relentlessly persisted until Tex finally gave in and jammed him into the small single seat front cockpit of his small low-powered Great Lakes trainer and taxied out for the take-off.

From this point let us turn the story over to Colonel Pettis:

"Well, Tex made it off all right with that big Indian crammed in there, but I wouldn't have given a thin dime for his chances of clearing the pine trees on the opposite side of the field. As it was, it looked as though he was mushing around between them and the last we saw of him he was still down among the trees.

"After about 15 minutes and Tex hadn't returned, we decided he hadn't made it and were about to go looking for him, when lo and behold! Here came Tex just a-kitin' in that little Great Lakes—but WITHOUT THE INDIAN!

"He landed and taxied up with that big grin of his on his face.

" 'Where's the Indian?' I asked. 'What did you do with him?'

" 'Oh, the Indian,' Tex replied, trying to look innocent. 'He's standing out in a field about ten miles from here.'

" 'Well, what happened?'

" 'Oh, after I staggered past those trees about one rev ahead of a stall, and still didn't gain any altitude, I just picked the first field I saw, sat 'er down, told that Injun to step out a minute and when he did I just gunned the ship and took off. Last time I saw him, he was still standing in the same spot with his mouth open.' "

87

IN PLOW LANGUAGE, "S" IS FOR SCRAM!"!

In the long-past glamorous days of flying, pilots for good or sneaky reasons plopped their aircraft down in grain fields, watermellon or strawberry patches and sometimes even in over-sized hen yards or pig styes. Now farmers, ranchers and the pork and fowl lot didn't cater to such sit-down shenanigans, although frequently the pilots couldn't help themselves due to mechanical failures. Other times, however, they DID help themselves—to the goodies in the patches or on the vines. Then there were the nice clear farm areas which made excellent spots for landing practice or from which to do a little passenger hauling.

Just such a clearing presented itself to pilot Charles McAllister and his student, Russell (Buss) Schlosstein on Sunday, April 21, 1928. They had started out from their headquarters at Yakima in a combination-wing OX5 Eaglerock to do a little barnstorming at Toppenish, Washington. This barnstorming business helped augment the meager earnings eked out of the flying school.

As the duo neared Toppenish, McAllister's eagle eye spotted a triangular-shaped field near the Northern Pacific Railroad tracks on the southeast side of the small town. It seemed a quite suitable site for their barnstorming operations. Their approach apparently had a dinner-bell effect upon a mass of citizens in that area who streamed in from every direction, beating the plane to the field. Since Buss had no trouble selling tickets to the eager customers, Charlie was kept on the hop.

In the midst of this up and down business, while Buss was loading two more passengers into the Eaglerock, Charlie noticed an Indian riding a white horse on the outskirts of the crowd. The Indian looked very unhappy; in fact a second look told Charlie the rider was down-right peeved. What Charlie didn't realize was that by landing here, he'd set his prop for an Indian uprising as this nice stretch of land was, in reality, now a war path!

Now the Indian was shouting in a deep booming voice that seemingly resounded the length and breadth of the valley and, of course, attracted the attention of all and sundry.

"No stop here—no stop here!" scolded the redskin, wildly waving one arm. Even without the verbal admonition, the angry Indian's flailing arm said it in sign language that needed no interpreter. Charlie could read him loud and clear—and in no uncertain Injun terminology!

Charlie thought a little heart to heart chat with the gent on the white steed would smooth his ruffled feathers; maybe Charlie could even get him to ride the aerial horses. Charlie therefore tried to wave the Indian over but he wouldn't come near. Charlie finally talked one of the young local spectators to go over to inform the Indian that they would pay him well for the use of his field. This also proved of no avail. The Indian simply resumed riding the outskirts and booming, "No stop here!"

Since Charlie and his student could see no valid reason for not "borrowing" this strip of land temporarily, they chose to ignore the redskin who continued to whoop it up every time the plane descended for another passenger. At the end of the day they took off for home port in Yakima, very happy over a most prosperous day.

The following Sunday they decided to go back with the hopes of repeating the success of the previous Sunday's "take." This time they found the field was full of horses. They were not to be defeated, however. Charlie hit upon a scheme to rid themselves of the "obstacles" in their landing path. He repeatedly swooped down over the horses until they were successful in getting the frightened cayuses to one end of the field so they could land. Then they hired some local lads to herd the horses all afternoon so barnstorming operations could be carried on without interruption.

During the afternoon the Indian appeared on his white horse several times at a distance but this time he emitted nary a sound. So, at the close of another successfull Sunday, the airmen headed for Yakima.

After those two fruitful Sundays, they decided to return to Toppenish for some more passenger-hauling but this time the lone redskin won the battle against the two flying palefaces. Instead of nags all over the place, Charlie and Buss saw something else—something that clearly announced their prosperity had come to an ignominious end. The Indian land-owner had plowed long "S"-shaped furrows over the entire field!

The Indian, apparently having had a serious pow-wow with himself, had conceived this wily scheme to out-smart these irritating palefaces with the noisy skybird.

THE DAY IT RAINED LUMBER IN BURBANK!

Meet Tony LeVier, the world's leading test pilot, who has been wringing out everything Lockheed has put wings on since 1941! Naturally a lot of things have happened in those thirty-plus years and the very fact that Tony is still with us today—and still "wringing 'em out" for Lockheed—is all the proof we need that he's the best there is!

Chief Lockheed test pilot, Tony LeVier, a great pilot and a great guy, who had many queezy moments.

One of Tony's "favorite" incidents—if there ever was such a thing as a "favorite incident"—was the time he plunked a Ventura bomber down in a Burbank lumber yard. That might be recalled as the day it rained two-by-fours in Burbank!

In January, 1942, Lockheed assigned Tony to run production flight tests on an advanced version of the Hudson bomber called the Ventura. The Ventura was similar in design to the Hudson, with the same wing and twin tail, but, overall, it was a larger, heavier craft with more powerful engines and "boo-koo" performance.

Ralph "Peewee" Virden, a former coast airmail pilot turned test pilot, had run the engineering tests on the Ventura under Rudy Thorsen's eagle-eye, which got the new bomber to the production test phase and into the clutches of LeVier.

Among the many things ground-and-air-checked in production testing was the landing gear warning horn. This is an important gadget that sounds like the Queen Mary arriving in New York harbor when the pilot forgets to lower the wheels for landing. It's also the gadget that, when blowing in the pilot's ear, keeps him from hearing the control tower tell him his landing gear isn't down.

LeVier had been flight testing the Ventura about a month when—oops! it happened! He'd just finished his horn test when his instruments told him all was not wheeling well in the wheel wells. This time the horn DIDN'T blow and that, in itself, was a blow! A look out the window showed him the right wheel was only part way down. Attempts to get the wheel the rest of the way down met with as much success as the guy who painted moth balls silver and tried to sell them to Pratt & Whitney for ball bearings.

Finally deciding the gear was up to stay, LeVier radioed the factory for advice. This, too, took some doing due to lack of direct radio contact with Lockheed flight operations. LeVier's message had to go through the airport control tower which, in turn, relayed it on to Lockheed flight operations—and vice versa.

At long last LeVier got the word that Pat McCarthy and Rudy Thoren

What's left of a Burbank lumber yard after being flattened by Tony LeVier & Company's Lockheed Ventura bomber.

were on their way up in a Lodestar to have a look-see at the situation.

Meanwhile Tony had the engines throttled back as far as he dared to preserve fuel. Even so he was fast approaching the proverbial belly landing he had in mind. His plan was to nurse it down under power until the fuselage touched the runway, then chop the throttles and skid to a stop. But the whole plan blew apart when operations told him to hang on a while longer.

And if "wait" broke the wagon, wait and see what happened to THIS sky-buggy!

By this time LeVier was practically skimming the rooftops. His fuselage tank was empty and he switched to his auxiliary wing tanks. Enter another bugaboo (they have a nasty habit of coming fast in situations like this!): In order to draw fuel from these wing tanks, the cutoff valve from the fuselage tank had to be closed. Would you believe the crew member assigned to perform this menial task flunked his job? He did! Instead of gas from the wing tanks, the fuselage tank began feeding vapor and air to the engines. This caused them to start cutting out and LeVier found it impossible to get fuel from the wing tanks.

By then he was flying due north at a point just east of the Burbank airport. From this point he could have turned and made it in okay. Realizing this, Tony alerted the tower for an emergency landing. This meant clear the decks of aircraft and vehicles and do it fast! He was under five-hundred feet and down to ninety miles an hour. The engines were coming in and going out in large surges of power, and Tony was fervently hoping the surges wouldn't zig when they should zag and throw him into a stall.

Just as he started his turn toward the field, the right engine began letting out mighty roars—then the left, and so on, until he was almost tossed into a spin in one direction then another. Tossing a fast look over his shoulder he discovered his flight mechanic, a throttle in each hand, pumping them back and forth as though he were operating a hand-car along a railroad track. Yelling at the mech to knock off the throttle exercise and get the flaps down, LeVier knocked both throttles into closed position. But by this time any chance of getting into Burbank airport had gone up in a mechanic's boner, for straight ahead of them, in the following disorder, was a flock of buildings, a mass of power lines and a bunch of revetments. The only sure thing they had to look forward to here was a nasty pile-up. Making a hard right to vacate that scene, LeVier spotted what he thought was an open field and instinctively knew he had the situation licked.

Drawing on all of his skill (he had a lot of it but a lot more shows up in tight squeezes like this), LeVier headed for that clearing like a bear for a honey tree, nursing that Ventura along for all it was worth to clear a last row of houses.

Just as he pulled the ship's nose up for the touch-down, he glanced ahead and almost fell out of his seat! Farther out on what he had thought was a beautiful clearing was a lumber yard with seemingly acres of lumber and box piles! He had been so intent nursing that bomber for survival that he had completely missed seeing the expanse of soon-to-be toothpicks! The only thing left to do was to try to steer his bomber down one of the aisles between the piles of lumber and hope they didn't anchor lumber piles down with chains! And down he went!

The first pile sheered off the left wheel and slued the ship slightly to the right. Then, with the left wing trying to imitate a surfboard on rough water, they porpoised up and down for seemingly endless miles although it was only a couple of hundred yards. The left engine, out in front, was acting like a battering ram with lumber and boxes flying in every direction. If ever there was an explosion in a box factory, it couldn't possibly have held a candle to the maelstrom of havoc LeVier and his crew were creating here. The ship was pitching like a raft in a Pacific storm, but at long last the bomber skidded to a halt amid an unbelievable shambles of lumber, boxes and what then should have been a vast silence! But was it?

No! That infernal horn had started blasting away indicating the landing gear wasn't down!

UNENDURABLE ENDURANCE!

Some situations could come under the heading of "Heroes—or heroines—are made, not born—and even that is dubious!" Here is one of them.

Ann Bohrer was one of the first three girls in Oregon to learn to fly; the first girl to land an airplane on Portland's then famous Swan Island Airport; the first girl in the state to make a parachute jump—and, if she hadn't run out of gas, probably would have been the first gal in the United States to set a women's endurance record!

According to Ann, it all happened thus:

"Portland, Oregon's Swan Island Airport was considered in 1927 to be the world's most beautiful airport and ever a mecca for week-end sightseers.

"Lured not only by the then still-novel aviation activity, they were equally attracted by the 'port's park-like beauty and its river-center setting immediately adjacent to the ocean-going channel. And, as a wonderful bonus, there was the Japanese cherry tree-lined drive completely around the 'port and its vast expanses of lush lawn and flowers. This week-end would tempt hundreds more because of a scheduled air show.

"Many of the student pilots were particularly nervous this day because of the closer-than-usual scrutiny of the many veteran air circus pilots lolling about awaiting the air show start. One ordinarily fine fledgling flier did such a loused-up job of flying and landing that his thoroughly exasperated instructor—and, unfortunately, MY instructor, also!—leaped out of the ship, stomped into his office, tore off his helmet and goggles and flung them into a corner. Then, glaring at me with blood in his eye, he barked:

" 'YOU take 'er up! Go out and show dose verdampte dumkopfs how it's done!' He was a quick-tempered German-born pilot who later became a well-known Alaskan bush pilot.

" 'Who, ME?' I asked weakly, fervently hoping I was mistaken.

" 'Yes, YOU! I vont a GIRL to show dem how to fly!' "

"I was appalled—in fact, 'shook up' better describes my feelings at the time. And there was good reason. Girl fliers attracted wide attention as they were still a rarity, or an oddity, and you can bet such attention wasn't always flattering, especially from some of the airport's male populace who resented having the field of aeronautics fouled up by females. The girls naturally had to withstand a goodly amount of jeering and teasing, and being the only pilotess on Swan Island Airport at that time, I came in for more than my share. Although I had become accustomed to the jeers, as well as pranks, of the airport gang, I had yet to get used to flying before an audience because of the inherent fear of being ridiculed for the slightest mistake and hearing, 'That's a girl for you!' I therefore flew around 6:00 a.m. when only a minimum of instructors and fellow students ventured to the field. And now to take 'er up, not only before the gang, but in full view of a few thousand visitors was absolutely out of the question.

"All eyes in the airport office were now watching me intently. 'But I've already BEEN up today,' I yelped. 'I flew early this morning!'

"The instructor's eyes were hot with fury. 'I don't care HOW many

They could have shot her down -- and shot her again afterward!
Ann Bohrer in her uniform as hostess of the Salem, Oregon Airport during the early 1930's.

times you've been up today—take 'er up AGAIN!'

"'But—,' I began, then stopped. The instructor had sailed out of the door. I knew I had lost the argument. The 'mech' looked at me with a smirk as he gathered the three extra pillows and four-inch rudder extensions, expertly fashioned for me by one of the United Air Lines' mechanics, which I required to see over the side of the ship and to give full rudder.

"He had never relished this extra chore as to him it only proved woman's place was on the ground and having to perform this accommodation twice in one day was just too much. I was sure he hoped I'd make a fool of myself.

"As he stalked off to the plane, I quickly pulled on my helmet, pushed my hair well under, and lowered the cuffs of my overalls to hide my shoes so as to appear more like a boy. I felt I just might get away with this subterfuge. As it turned out, I didn't get away with anything!

"When the mechanic climbed out of the plane, I made a wild dash across the apron and into the cockpit so as not to attract any more attention

95

than necessary. As I ruddered around to taxi onto the strip, the owner of the craft ran after me. 'I forgot to tell you,' he shouted, 'you've got only 10 minutes of gas left so get down before then!'

"The admonishment didn't perturb me in the least. The thing uppermost in my mind was did anyone suspect I was a girl?

"As I winged off, I decided one time around was enough to 'show 'em.' As I completed the sample run, I banked around the old Portland flouring mill and throttled back for the descent. It was then I noted a plane taxiing on the runway so I circled the field again but by the time I had gotten lined up with the runway, another plane taxied out. Not knowing how long it would be before the craft would take off, I decided to make another circle flight. At the end of the third 'round' the runway was clear but just as I started my glide, a plane with more 'horses' than my 90-horsepower OX5 darted in from beneath me to make a landing. It occurred to me I should have attached a banner or streamer to one of the struts to indicate a student was at the controls and perhaps these experienced pilots would have accorded me the right of way, which was the usual procedure. Such streamer wasn't necessary at 6:00 a.m.; therefore I never used one and I didn't think of tying one to the plane now, nor did anyone else. In the circumstances, none of the pilots were going to show me any special courtesies. It was every man for himself, there being no control towers in those days.

"As I circled about once more, I became somewhat worried. It was the first time I had given the fuel supply any thought. How much longer was this 'ring around the rosy' going to last? I recalled the admonition of having only 10 minutes of gas left. This time I would HAVE to come in!

"I swung around for the landing approach after first checking if any other ships were in close proximity. The sky was clear in the immediate vicinity but alas, once again the landing strip was not! Another pilot had beaten me to it. If I were sure the fellow would stay put, I could fly in over him but I wasn't about to tempt fate. To me it was particularly important that I make an excellent landing—not only to 'show 'em,' but because I was a girl and I may not have fooled anyone into believing otherwise. I later learned that as soon as I had gotten into the plane, news spread through the crowd like wildfire that a girl was about to take-off. From there on, all eyes were riveted on my plane, especially when they noted the difficulty I was having in coming in to land—and almost out of gas, yet! I think they felt they just might witness a nice crash-landing. I was indeed the 'star' of the show! My brother, Walt, who was told I was the one who was attracting all the attention, joined the other airmen on the apron who'd excitedly shout, 'NOW you can come in—everything's clear!'—as though I could hear one measly word! Somehow my fellow skymen overlooked the fact my 90 'horses' just couldn't gallop around quickly enough to take advantage of the 'clearings.'

"The 10 minutes I knew were gone. Those below also knew and I was sure they were becoming extremely uneasy as they intently listened for the inevitable 'sputt-sputt' which would herald a dry tank.

"As I went around again, I knew this would have to be it but it looked as though I might not even make this one. Were it not for the loungers on the grass and the sprinkling system, I could land on the lawn although airport rules prohibited lawn-landings because of the tail skids tearing up the turf. If I broke the rule, it would mean another run-in with the airport 'cop,' one Ferdinand Gauntt, whose wrath I'd faced on too many previous occasions when the lawn invited landings and I thought nosey old Ferdinand wasn't looking! And he certainly was looking today and, as usual, wouldn't excuse the fact that I usually made two-wheel landings and so avoided any tail skid marks.

"I was now aligned for descent and the landing area awaited. A hasty check of the sky around me indicated all was clear. Just as I closed my throttle I spotted a yellow something directly beneath my wings. Leaning far over the side I received the shock of my life! A speedy craft had overtaken me and slipped into a blind spot beneath me. Giving full throttle, I hastily pulled her up and away. That was a close one! By now I was more aggravated than perturbed. Since I had to land into the wind and time was of the essence, to put it mildly, I made another turn—a tight one—around the field. I HAD to get down before anyone else cut in. Checking right and left, fore and aft and underneath, I found all was clear. I closed my throttle and made for the runway which now had been cleared for me.

"As I swooshed down, I noted that wasn't all they'd done. On the apron, ready to spring into action, were a fire engine and an ambulance in all their red and white glory! They were all set for a 'dead stick' landing with possible disastrous results. All faces were turned toward me. I had to smile. I was sure all were very concerned and nervous and here I was—cool as a cucumber! Overcoming the urge to chase those loungers off the soft lawn and plow through the whirling sprinklers, I made my usual 'hot' two-pointer on the gravel runway and taxied to the apron. As I reached to cut the switch, came the long-expected 'sputt-sputt'—out of gas!—and along with it also came the owner of the plane, also 'sputt-sputtering!' He was followed by others who'd been mentally trying to get me down these 15 or more minutes.

"The owner's face was livid. 'Vot were you trying to do?' he roared, 'BREAK THE WORLD'S ENDURANCE RECORD?' With that he turned on his heels and stalked off.

" 'Well, you could SEE I couldn't get down!' I yelled after him but the man was too furious to make reply. Now for the first time since my take-off, I became upset and the surging crowd offering congratulations didn't ease the guilty feeling that perhaps there was something I should have done and didn't know about. After all, I was just a new soloee.

"It was the next morning when I wanted to fly that I learned punishment had been meted out. I was to be 'grounded' for 30 days—the 30 days the aircraft owner knew I sorely needed to practice as at the end of that time I was scheduled to face the Department of Commerce inspector for my private pilot's license.

"It was indeed an endurance contest for which I didn't win any laurels!"

But look what happened to Ann a year or two later. By that time Ann actually DID make arrangements to attempt to break the world's endurance record for women. In this connection a specially-constructed plane was built for her through the sponsorhsip of Rolla Simmons, noted Portland restaurateur, whose restaurants she was to advertise. However, this time, ironically, the *plane* was grounded! It couldn't pass F.A.A.'s rigid standards!

THEY SHOULD HAVE YELLED "FORE!"

Golf course fairways have ever been the pride of their keepers, to say nothing of the country club owners, and in the old days of tail skids, airplanes on golf courses were about as welcome as the proverbial skunk at a lawn party.

One of the finer golf courses in the state of Washington in 1928 had the unfortunate circumstance of being situated on a route plotted betwixt Yakima and Everett, Washington, by three or four fliers returning to Everett from the huge air show dedicating the brand new Yakima Airport.

The Langdon brothers, John and Warren, of the Everett Air Service were flying a Ryan Brougham, a sister-ship of Lindbergh's "Spirit of St. Louis," sporting the spritely name of "Skylark of Everett." Approaching Ellensburg at 3,000 feet, the usually very reliable Wright "Whirlwind" engine suddenly turned into a *Wrong* "Whirlwind" by sheering a cam shaft gear, necessitating a forced landing. At this point what more welcome sight could there have been than a beautiful green golf course spread before them so down they went!

Pilot Johnnie Langdon gives a happy thought
to his 500-foot long "divot"!

Landing on the fairway directly in front of the country club headquarters, the tail skid of the heavy Ryan plowed a "beautiful" furrow six inches deep for a distance of 500 feet down the loveliest groomed green it had ever been their pleasure to behold! Coming to a stop, it wasn't long before the greenskeeper, the president of the country club and other lesser fry descended upon them in anything but jovial welcoming moods.

"Look at what you did to our fairway—you've ruined it!" shouted the greenskeeper.

"You'll pay for this!" shouted the president of the golf club.

The lesser fry added other similar endearing terms while Warren Langdon tried in vain to explain that when an airplane engine stops running, there is but one way to go—down.

About this time another motor was heard. Looking up, here came another ship circling the links—an OX5 Travel Air biplane piloted by one inquisitive "Dinty" Moore. And as the faces of their golf course "welcoming comittee" ran the gamut of pure shock, rage and sheer resignation, in came "Dinty," plowing another furrow paralleling the one the Ryan had just made!

"What the hell is this, a g—d—ed air show?"

Of course insult was added to injury when a new rope was borrowed with which to tie the Ryan down for the night, and pilot Johnnie Langdon calmly proceeded to chop it in half with a hatchet!

"Ya gotta have a rope for each end of the wing, don'tcha?"—an apologetic question asked as though it would clear up the whole matter.

Somehow it didn't.

IS THIS PILOT NECESSARY?

And it is things like the following that make you wonder at times whether pilots are really necessary.

A southern California pilot, one Lieutenant Knud von Clauson-Kaas, one time so-called dare devil of the Danish Royal Air Force and later known to his friends as Denmark's gift to commercial aviation, wanted to try out a new Golden Eaglet monoplane in 1929 at Los Angeles. However, distributor W.G. Mead was somewhat dubious about the lieutenant's ability, so he went along in the front cockpit. But Clauson-Kaas, who had some friends at the field who had never seen him fly, told them to watch this flight closely as he would be at his very best.

He climbed in, warmed the motor, and gave it just enough gas to start down the field with the tail skid off the ground. Half way down the motor opened up with a roar and the ship leaped ahead. One wheel left the ground and then the other. The plane began to skid to the left, making a little altitude but barely clearing the last hangar. Then she wobbled a couple of times, did a dip, and made a turn that brought it face to face with a gas tank on the left side of the field. By this time Clauson-Kaas' friends, who were thinking that maybe he wasn't such a hot pilot after all, began to wonder how safe they were with a gas tank explosion due in a few seconds.

But the plane cleared the tank with another skid, climbed a few more feet, and circled back of Western Avenue and came down for a good landing. Then it taxied back to where the friends were standing and Mead hopped out of the cockpit with blood in his eyes.

"What the hell!" he shouted. "Who told you you could fly?"

"Me?" countered the Dane. "I didn't TOUCH the stick—I thought YOU were flying!"

It seemed Clauson-Kaas had started the ship down the field and Mead had accidentally bumped the front-cockpit throttle wide open. At that Clauson-Kaas figured Mead had taken over and took his own feet and hands off the controls. After the gas tank episode, Meade grabbed the stick and brought the Eaglet in.

And in all of his 20,000-plus-hour flying career, Harold Bromley's funniest flying experience was a perfect match to the preceding story.

It happened during the 1930's. Bromley, then an Inspector for the U.S. Department of Commerce (equivalent to today's Civil Aeronatutics Administration), was flying with Art Jenks, a well-known Modesto, California, flight instructor. In a Piper Cub at 2,000 feet, Bromley would do a maneuver and then Jenks would take over and follow through.

They had just completed a 720-degree turn. Bromley wobbled the stick in the dual-controlled, tandem-seated Piper, indicating that Jenks was to take over the controls. For some unfathomable reason, and unbeknownst to Bromley, Jenks failed to "get the message"—probably thinking of the beautiful blonde he eventually married. They gradually lost altitude and the plane went into a spiral, getting tighter all the time.

Bromley, who naturally assumed Jenks had the controls, was getting a bit worried at the proximity of the power lines below them and the speed with which they were whirling up. He finally shouted to Jenks in the front seat:

"What the hell are you trying to prove?"

Art yelled, "Hell, you have the controls! I am wondering the same!"

Needless to say, they unwound in a hurry and landed just in time to keep the City of Modesto from "enjoying" an electrical blackout and the Modesto Bee from having lurid headlines!

SPEAK PLAINER, PLEASE!

Misinterpretation of pilot instructions quite often has been the cause of passenger panic. A case in point is the experience of Dick Depew.

Depew had been hired by a salesman to fly him across the state of Pennsylvania on business. Since the weather was bad and the forecast nothing to cheer about, Dick had equipped himself and his passenger with parachutes, with instructions to his passenger on how to use it.

"Chances are we'll never need these," explained Depew, "but IF we do, I'll just yell 'Jump!' and you bail out that door, count three and pull the ripcord!"

The salesman nodded a most unenthusiastic agreement, accentuated by a set of very white knuckles.

The weather progressively worsened and the ship was being buffeted about so hard that Depew was having difficulty staying right side up. Simultaneously his engine commenced cutting out and Depew realized the turbulence was affecting the gas flow to the engine. Now, as some pilots did, Depew had installed a small "wobble pump" down to the left and toward the back of the seat. With a little pumping help from his passenger, the gas flow problem could be alleviated, so he yelled, "Pump! Pump!" to the salesman.

A sudden roar of wind filled the cabin and Depew glanced around—just in time to see the salesman disappear out the door! His instruction to "Pump! Pump!" had been misinterpreted as "Jump! Jump!" And he did!

The salesman's 'chute opened and he landed okay except for a few bruises and a sudden urge to travel by train.

Another example of passengers misunderstanding the pilot occured during a routine passenger-hauling flight over Yakima, Washington, in an open-cockpit Waco biplane flown by Charles McAllister of Yakima's McAllister Flying Service.

As was McAllister's custom when carrying passengers on sightseeing flights, he throttled back and pulled the nose of his ship up slightly in order to fly as slowly as possible. This way his passengers would have a less windy, more enjoyable view of the country side. Before realizing it, he noticed the propeller had stopped turning completely. Since he was certain there was nothing radically wrong with the engine, other than idling too slow, he decided to dive the plane almost straight down for a thousand feet to restart the engine, there being no self-starters in those open-cockpit days.

Being a courteous pilot, however, McAllister sought to reassure his passengers, so he called out as loudly as he could: "We're going to dive!" Immediately his passengers assumed the most terrible expression he had ever seen on anyone's face before or since.

After the dive, with the engine again running smoothly, he landed and taxied back to the hangar. There his still ashen-faced passengers told him they had understood him to say they were going to "die" instead of "dive!"

THE COW SHOULD'VE BLOWN ITS HORN!

During those early barnstorming years, pilots were forever locking horns with the lowly bovine. Many a farmer's or dairyman's wrath has been incurred by aviators "doing in" their Jerseys with their Jennies. But have you ever heard of a cow being neatly dehorned by an airplane?

It has happened!

A Texas pilot, Joe Palmer, accomplished the trick some years ago at Mexia, Texas. While gliding in for a landing, Palmer discovered a cow quietly grazing in the path of his descending Curtiss Robin and zoomed up sharply to avoid parking his ship on the bovine's back. His propeller dehorned the animal slicker than a whistle, at the same time also "dehorning" the tips of his wooden prop.

He made it into a nearby cotton patch without further damage other than coming face-to-face with an irate horny-fisted son of the soil who demanded immediate reparations for one dehorned cow and a half acre of damaged cotton.

Palmer agreed to settle—why, we'll never know!

GHOST RIDER IN THE SKY!

Glider pilots, too, came in for their share of humor while soaring about in their motorless craft. Any glider pilot will tell you that in the solitude of noiseless flight it often is possible to hear train whistles, bells, or even conversations on the ground. By the same token it often is possible for those on the ground to hear a glider pilot calling to them—and this they often did without knowing where the voice was coming from.

Such a situation helped set the stage for a humorous incident experienced by Yakima, Washington, pilot Charles McAllister who, in earlier years, was an avid gliding enthusiast. Let us herewith allow McAllister to tell his own story:

"It must have been about 1932 when we began launching the 'Yakima Clipper,' the glider my brother and I had taken two years to build. I well recall the particular incident I am about to relate as I had made a 'shock cord' take-off from the Yakima Ridge at a location called Coyote Springs, roughly five miles northeast of Yakima.

"About a dozen or more Yakima people were present at the launch side and they, of course, eagerly helped to launch the 50-foot-wing span glider. I was at the controls, and after a nice start, spent about 15 minutes making the typical "S" turns back and forth on the windward side of the ridge. A mild rain squall was in progress so the lift was very good since the wind was blowing some 40 miles an hour. I began cruising westerly along the ridge across the Yakima River at Selah Gap, then past a place called Lookout

Yakima, Washington, pilot Charles McAllister and his sailplane, "Yakima Clipper". On June 16, 1933 McAllister stayed aloft 18 hours, 33 minutes in this motorless craft.

105

Point. This is a popular spot on the 600-foot ridge that commands a wonderful view of the city of Yakima looking south with the Naches River below. Practically every young man who has grown up in the area has taken his girl on a 'sight-seeing' trip to Lookout Point at least once.

"After cruising the six to eight miles from Nelson's Bridge on the Naches back to the launching site, about a 15-minute flight each way, I noticed a Ford coupe parked on the Lookout Point turnaround. Since it was still raining slightly, there was no other traffic on the road. I had been flying some 1,500 feet high so I decided to come down to about 100 feet over the parked coupe on one of my round-trip circuits along the ridge and give the people the thrill of seeing the 50-foot wingspread glider. Since at about only 30 miles per hour, the glider makes almost no noise, the people in the car could not hear it so I called out, 'Hello! Hello!' as loudly as I could. Immediately the two occupants got out of the car—a fellow on one side and a girl on the other—and looked in every possible direction but up. They then got back in their car.

"Fifteen minutes later, after making the circuit to Coyote Springs and back, the car was still there, so I repeated the low 100-foot sweep and again called out, 'Hello! Hello!' Again the young couple piled out of their car, this time a bit faster. The boy then walked to the south side of the ridge and his companion to the north side, and again they looked every which way but up. They apparently never did see the glider but seemed greatly disturbed over the source of the eerie voice out of 'nowhere.'

"After another circuit east on the ridge, I returned in time to spot that Ford coupe leaving an unusually fast trail of muck in its wake as it skidded down the grade away from that 'haunted' scene. I continued my flight, chuckling over my apparent role as 'Ghost Rider In The Sky!'"

THE CURRENT PROBLEMS OF A "BALD EAGLE!"

How to keep a whole city in an uproar is a topic on which one Pacific Northwest pilot should be an expert. He is Art Whitaker, better known in flying circles as the "Bald Eagle of Aviation."

Art's electrifying—or un-electrifying—tale of woe happened at the Salem, Oregon, airport while he was giving a student flying instruction in a Curtiss Robin monoplane. This was during the time that a few local air enthusiasts were towing a glider off the airport with a car, using a long, heavy wire as a tow "rope." This may have been all right except for the fact that they had a nasty habit of leaving the wire lay on the field.

Whitaker usually liked to take off toward the southwest right from the ramp in front of the hanger—something the Federal Aviation Agency might frown on today but at that time pilots had more important things to worry about, like food, for instance.

The glider wire must have been lying north and south because it was immediately snagged by the tail skid of the Curtiss Robin. However it made no noticeable difference in the flying performance of the ship and so went unnoticed by Whitaker and his protege. Neither could they see the wire because there was a lot of privacy in the Robin cabin, especially the rear seat.

They shot landings (made touch-and-go landings) for about 45 minutes and didn't have a bit of trouble getting the tail down! About that time they

The Pacific Northwest's "bald eagle" of aviation, Art Whitaker, the culprit responsible for the great Salem "blackout" of the thirties!

noticed the whole airport crew out on the field, all seemingly putting on some kind of an act. Some were dancing up and down waving, some making hand signals as though attempting to land a plane on a carrier, and one was all bent over waving one hand between his legs and Whitaker thought he had decided he was one of Tarzan's apes. But then one could expect most anything from that Eyerly Aircraft gang!

When the last landing was finally made and Whitaker and his student taxied back to the hangar amid the still-cavorting ground crew, it developed that the electric power had not only been going on and off in the Eyerly shops but in the ENTIRE CITY OF SALEM! The whole city powerplant crew had been tearing up and down the panels as fast as they could, pushing in circuit breakers. But as fast as they got them set, the old Robin dragged the wire over the main power lines and they all blew out again! This happened every three minutes for the best part of an hour so the Salem powerplant crew had one of the most active hours in their usually sedate careers attempting to keep the city in power—something like "the mail must go through!"

TALLMAN———UNFURLED!

You can take it as gospel from motion picture pilot Frank Tallman that hanging on to the upper wing of a Navy training plane in a near vertical dive, your body flapping out into space like a windsock in an Alaska williwaw, with the wind whipping your clothes off faster than an exterminator with a pants full of hornets, isn't exactly the most comfortable position on or off the earth in which to be!

Nevertheless, that is precisely the ridiculous situation in which Tallman, cohort of the late renowned Paul Mantz, found himself one ten-degree day over Glenview Naval Air Station, Illinois.

In addition to being able to flap in mid-air like a battle flag, motion picture pilot Frank Tallman is licensed to fly everything that zooms, bounces, flies or crawls into the air, except rockets. This includes single and multi-engine airplanes, 'copters, gliders and balloons. Former Navy fighter pilot and instructor, Tallman, as president of Tallmantz Aviation, Orange County Airport, Santa Ana, Calif., now furnishes the movie industry with aircraft of all types and ages that zoom, fly, crawl or bounce into the air.

But who better than Tallman himself is best able to unfold such a tale? The answer being quite obvious, we allow Tallman herewith to take the controls:

"My funniest flying experience occured while I was a Navy officer and flight instructor at Glenview NAS late in 1942.

"It was winter and cold enough for the proverbial brass tree swinger. I was flying a Naval aircraft factory N3N3 open-cockpit biplane and had a ham-fisted cadet in his spin stage—which, to the uninitiated, means that I was attempting to teach him how to get out of a tail spin should he ever happen to find himself in such a predicament. This was before the days of shoulder harnesses, and my problem, which follows, was—I understand—in part responsible for the Navy retiring the old, wide lap belts we had then, and which depended solely on the over-center principle of locking.

"Even with the Navy's best fur-lined flying suits and woolens enough to keep Pendleton happy, both the student and I were frozen. The student froze figuratively and literally on the control stick during the spins, and I would have to take over and labor like Hercules to get him off it before we spun in. Each and every spin was followed by the necessity of me taking over. No amount of cajoling over the speaking tube could seem to make the student dump the stick forward and kick opposite rudder—the control requirements to get out of a spin.

"I was sizzling in spite of the frigid weather, but, like all good instructors, I stayed off the controls to give the student the knowledge that HE was doing the flying—not I; only taking over at the last moment when the student couldn't recover.

"On the fifth spin we started at five thousand feet and, after two turns, I told him to bring 'er out. At that moment I reached across my waist with my fur gauntlet to take over as I had to do on all previous spins. But, unbeknownst to me, the fur cuff snagged the lap belt plate and moved it past dead center. Also at that precise moment, all of my past proddings via the speaking tube suddenly hit 'home' in one fell swoop, causing one heavily-muscled Iowa farm boy to stab the stick forward like a javelin!

"Now in a true weightless state and experiencing an astronaut's training a score of years ahead of any space program, I sailed out of the cockpit—no longer hampered by a seat belt—like a China Clipper departing Hong Kong with all sails drawing!

"Despite this surprise ejection, as I passed the upper wing I grabbed the hand grips and promptly trailed out into the slipstream like a battle flag!

"With the ship now in a near vertical dive with the speed building up alarmingly, I was shedding clothing like a Salvation Army truck going down a rough hill! All of my hollering, before the speaking tube departed for parts unknown in the air blast along with my helmet and goggles, did no good because in my flying completely out of the cockpit, it had disconnected.

"Afraid that the student would again freeze on the controls in the vertical dive, I desperately fought the air blast enough to kick a boot through the windshield, then hook the cockpit rim, then the instrument panel, and finally—whew!—I was back in the seat, and none too soon! With the ground coming up much too fast for comfort, frozen stiff, looking like a scarecrow and madder than a wet hen, I grabbed the contols and yanked the ship out in a zoom that would have made a steel bridge bend a girder. I next came out of the resultant blackout in a vertical attitude just entering a tail slide!

"We got back to the air station okay—tho' the good Lord only knows how—myself dressed as though for a Turkish bath rather than for flying in ten-degree temperatures.

"The student's answer of 'I'll be darned if I know!' when asked what he would have done had I blown away like a maple leaf, was promtly followed by the only down-check I ever gave as a Navy primary flight instructor! I then went in and got on some clothes before being arrested for indecent exposure, or worse—like freezing to death!"

"IRON HAT," BARLINGS AND BUZZARDS!

When Walter Barling of World War I bomber fame came up with the idea of his little NB-3 low-wing monoplane in the late 1920's he had anything in mind but one "Iron Hat" Johnson and his fantastic flying foray into the state of Arkansas.

Along with such post-World War I aerial rip-snorters as "Speed" Holman, Tex Rankin, Al Williams, Gordon Mackey and others of their ilk too numerous to mention, "Iron Hat" Johnson, who arrived on this earth as Forrest Myrten Johnston, was an aerial spell-binder whose aerial capers spell-bound far more than his fair share of air circus gawkers during the years when flying was purely a seat-of-the-pants operation. He came by his monicker "Iron Hat" because of the hard English derby hat that perennially graced his cranium during his avionic escapades. His picturesque nickname and Chaplinesque appearance was a combination that always drew a substantial air show gate wherever he was programmed.

"IRON HAT" JOHNSON
America's Thrill Rival to Falconi and Udet

One bright day in late 1929 found "Iron Hat" together with a passenger, Chic Fisher, arriving on the ground at the little town of Camden in Arkansas. This feat had only been accomplished after the third pass necessitated by an armada of belligerent horses, cows, goats, and a razorback or two that insisted on defending their pasture rights. They had now

112

withdrawn, obviously to regroup for the next attack.

Johnson and Fisher had covered the 70-or-so miles from Glenwood, their point of take-off, in something like one hour, having been chased most of the way by Arkansas buzzards that apparently mistook the orange and black Barling for another buzzard. For those unfamiliar with Arkansas buzzards, let us say that these birds fly thousands of feet in the air, soaring hours at a time, some even going to sleep while soaring. In this fashion they log hundreds of air hours, but always keeping an eye on their nests. Some are so lazy they lay their eggs in mid-air, but sharp of eye they always manage to hit their respective nests like bombing a target. They never miss.

Having taxied to a standstill in a cloud of dust, Fisher crawled out of the cockpit and down off the wing to go in search of a taxi or other form of motorized ground transportation. He made off toward a would-be house built on stilts, with nary a window, with the hopeless thought in mind that the owner may possess a telephone.

"Iron Hat," meanwhile, had elected to remain with the ship, knowing from sad experience that a pilot cannot trust a horse, cow, goat—or razorback—around an airplane, since they either lick it, kick it, butt it, or try to eat it. He had left the motor ticking over hoping that the sound and the whirling propeller would keep the pasture denizens at a discreet distance, since they had now managed to form a large circle around the ship which they were slowly but surely tightening by the second.

Johnson stood this creeping encroachment as long as he could. No four-legged Arkansas bag-of-bones was going to eye HIS airplane as though reading a menu! With this in mind, out he hopped with the firm intention of shooing the entire pastoral menagerie into the next county.

He had barely commenced his shrill shooing when amid the accompanying moos, neighs, oinks and baaas of his reluctantly retreating audience, another more ominous sound assailed his ears—the roar of an airplane!—HIS Barling #9320 with not even a buzzard in the cockpit! The throttle had crept open and the craft was enroute to who knows where!

Heretofore, on demonstrations, when Johnson was attempting to sell a Barling to a prospective customer, he would always say, "This airplane will fly by itself!" but THIS wasn't exactly what he had in mind. This "Look, mamma! No hands!" stuff wasn't at all funny with him out of the cockpit. He could see the headlines now:

BIG MYSTERY! AIRPLANE LANDS IN TEXAS WITHOUT PILOT!

With the ship now headed his way, "Iron Hat" figured if he could grab the wing and spin the airplane around, MAYBE he could get to the cockpit and yank the throttle shut.

As it roared toward him, Johnson leaped toward the end of the wing, momentarily feeling some comfort in the thought that the propeller didn't reduce him to hamburger patties. Just as a semi-humorous thought crossed his mind—"One manpower overcomes ninety horsepower!"—he took the full impact of the wing in the general vicinity of his rib-cage!

When he came to, he was lying flat on his back looking up at a bright blue—though star-flecked-sky and instinctively knew his name had been scratched from the list of 100 most important people on earth, except for a list the buzzards might be carrying.

The airplane had vanished completely. So had the pigs, cows and other animals. Even Chic Fisher was gone. Could he be in Heaven? Nope! In all that dust he knew he still had to be in Arkansas. But he never felt so alone—or misused—in all his life!

After slapping half the dust in the state off his clothes, he headed for the windowless house on stilts to try to find out what had happened to Chic, and also to see if his airplane had gone "that-a-way." No one there. Chic obviously was on a continuing search for a phone. Just then a car went by and disappeared hell bent into a nearby woods. Then another. Johnson headed for the road, thinking to get a lift into town where he could report the missing monoplane to the local constabulary, and, in turn, have them alert Missouri, Texas, Oklahoma and Louisiana to get the children indoors.

Suddenly he spotted the cars at the end of the road with the occupants all looking into a brush pile. There, wrapped up in seeming miles of fence wire, was the remains of his Barling at the foot of a tall tree, with everyone poking around for the remains of the pilot.

Their first reaction at finally spotting "Iron Hat" alive within their ranks was "Where did you-all come from?" "What happened to this heah airyplane?" and why was the NB-3 all wrapped up like a Christmas package with fence wire for a ribbon?

What actually happened is pure conjecture. As near as Johnson could put two and two together, here is what may have taken place:

In bracing himself and absorbing the impact of the wing during an otherwise perfect—though pilotless—take-off down the center of the field, the Barling was deflected to the left and headed for Oklahoma instead of down the runway toward Missouri. In the process of gaining altitude, either the landing gear hung too low or a large roll of fence wire, left near the path of take-off by some farmer, was too high. In any event it managed to foul the landing gear and strung out behind the airplane like one of those rope ladders used by aerial stuntmen when changing from one airplane to another. Next a small tree practically emulated the part of an aerial stuntman by glomming onto the fence wire. The irresistible then met the immovable when a much larger tree took over from the small one. The Barling finally ran out of steam and altitude about half way up the larger tree with branches and pieces of the propeller sailing in all directions. Since a piston-engine airplane can't fly without a propeller, and the Barling now had no visible means of support, it and its cargo of wire fell out of the tree into the brush around which the now thoroughly amazed local citizenry was currently gathered.

The story is still being circulated in that section of Arkansas today—with variations, of course—of how one pilot was outwitted by his own airplane, or hexed by Arkansas buzzards and a field full of farm varmints. Some natives even claim one of the buzzards laid an egg aloft which hit a crosswind on the way down, causing the egg to deflect and strike the throttle of the Barling!

But who knows for sure? Only the "Shadow" knows!

THE "13 HOLLYWOOD BLACK CATS"

They looped Pasadena's Colorado Street bridge every time someone made a suicide jump from it; they played poker sitting on the top wing; they hung by their hair from the landing gear; they rode the top wing while they looped; they made car to plane and plane to plane changes—other than that, the rest of the things they did couldn't be done! Their slogan was "We'll do anything for a price!" and they wore black sweaters with a sneaky-looking black cat and the number "13" in a white circle across the chest—and NO book of this nature would be complete without chronicling a few of their aerial escapades!

These were the world-famous, cool-nerved "Thirteen Black Cats" of Hollywood who, for a number of years starting in 1924, thrilled the pants off movie-goers, air show crowds and even sporting event spectators as you shall see.

Notwithstanding such other "flying circuses" of the era as the noted "Gates Flying Circus" and the "Inman Brothers Flying Circus," the "Thirteen Hollywood Black Cats," through their world-wide motion picture antics via Fox Movietone, Pathe and International Newsreels, were far and away the most renowned—and the wildest group of them all— the "wildest" simply because everything they ever did was considered either way out of the ordinary or altogether impossible!

Even the way the group got started was out of the ordinary, but let's just turn the story over to Ken "Fronty" Nichols, one of the three co-founders, for the telling!

"The 'Thirteen Hollywood Black Cats' started back in 1924 when 'Spider' Matlock, 'Bon' MacDougall and I—'Fronty' Nichols—were interested in the old Burdette Airport out about 104th and Western Avenue in Los Angeles. We pooled all our nickels, corralled a few old army 'Hisso' Jenny's ('Hisso' for the 180-horsepower Hispano-Suiza engine and 'Jenny' for Curtiss JN) and dolled 'em up, figuring a few people, at least, were flying-minded and that more and more were bound to get that way. We three had been flying for some time and had wild dreams about the future of aviation—and, as it turned out, not nearly wild enough!

"Our idea at the field was to teach air-minded gals and guys all we knew about piloting and, as a side-line, sell rides to the people who wanted a thrill at five, ten, and twenty-five bucks a ride, dependant, of course, on the length of the ride and what we did during it.

"Sunday was our big day there, so, to bring in the crowds, we hired a couple of daredevil stunt men (guys as rare in those days as a buttercup on an iceberg) to come in and do some wing-walking and other didos on the ship. We shot our wad on a big ballyhoo for them, but when the great day came—they didn't! There we sat with pilots, planes, a bang-up Sunday crowd milling around waiting for something to happen, and not a single 'death-defying daredevil of the clouds' in sight! Were our faces—and our bankrolls—red! We had to deliver something fancy in the air RIGHT NOW, or give up our dreams and go back to the farm. I guess 'Spider' and I got the same idea at the same time. 'I will if you will,' he said. I took a deep breath. 'It's a deal, chum!' And without ever having thought of stunting in our lives, we took off with 'Bon' at the controls and, heaven only knows how, managed some plain and fancy wing-walking. The crowd loved it, and when the first shock was over, so did we!

Here they are! the daring young "cats" on their flying machine. Ken Nichols on the left wing and "Spider" Matlock on the right wing are grabbing hats from Paul Richter, on the ground left, and Art Goebel on the ground right.

"After that we decided this stunting racket was a cinch, so we proceeded to get together a gang of our own, figuring we might as well save the dough we'd have to spend for so-called 'professionals.' To begin with there were 'Bon,' 'Spider' and myself. We then rounded up Arthur C. Goebel, the pilot who later won the Dole Oakland to Honolulu trans-Pacific Air Race; Paul Richter, Jr; Ivan 'Bugs' Unger; Frank Lockhart; Albert Johnson; Herd McClellan; 'Wild Billy' Lind; Samm Greenwall, the International Newsreel cameraman; Morrison Stapp, and Reginald Denny, the movie actor. Denny was a swell pilot, a veteran aviator of World War I, but he was chiefly an honorary member of the Black Cats because the studio wouldn't let him do much stunt work with us. Unger had gotten into the game by doing free balloon parachute jumping with his Uncle Ed, the grand-daddy of all balloon jumpers. McClellan was an old-time circus stunt man and parachute jumper, and Lockhart was a racing driver. We had a dash of everything!

"We called ourselves the 'Thirteen Black Cats' and began to get a lot of movie work. We wore black sweaters with a big white circle on the chest inside of which was a black cat and the number 13. Everyone had to have thirteen letters in his name, or else juggle the spelling to make it come out to that number—count 'em! That's why Sam is spelled Samm in Greenwall's name! Even on the ground we leaned over backwards to defy every superstition we could find. Every Friday the 13th, we'd perform our newest, craziest gags for the newsreels. Naturally we got a terrific amount of publicity from motion pictures and newspapers. 'Us Cats' had just about cornered all of the moving picture and newsreel stunt business!

"In the late 1920's, the era of those Hairbreadth Harry and Helen escape pictures, and in the early 1930's when the World War I cycle of flying pictures began with a bunch of 'Air Devil' serials, and the great feature pictures, 'Wings' and 'Hell's Angels'—we were the boys who did most of the stuff you used to choke on your lollypops watching. And after some of the close calls we had, we decided that the name 'Black Cats' was perfect for us since we sure seemed to have the proverbial 'nine lives!'

"On October 31st, 1926, the following item appeared in the LOS ANGELES TIMES:

> Three young 'flying fools' in an airplane menaced the lives of 79,000 spectators yesterday between halves of the Stanford-USC football game at the Coliseum.
>
> The three, pilot 'Bon' MacDougall, 'Fronty' Nichols and 'Spider' Matlock, all members of the 'Flying Black Cats,' began stunting over the Coliseum during the half-time entertainment.
>
> As thousands shrank back in their seats, the plane swooped low with Matlock and Nichols standing on the wings.
>
> It barely cleared the peristyle end of the stadium.
>
> Spectators in the topmost seats were so close to the plane, they could distinguish the features of the fliers.

Pilot "Bon" MacDougal doesn't mind his "pedestrians" on the top wing! They are, from left, "Spider" Matlock, Al Johnson and Ken Nichols.

"I darn well recall that day! SC had asked us to give the crowd a thrill, but we didn't intend to make it that thrilling! We had made elaborate preparations for the stunt, even to footballs painted in USC and Stanford colors and parachutes to match, hand-made by my wife. As we approached the Coliseum, 'Spider' and I climbed out on the ends of the upper wing of MacDougall's plane, stuffing the footballs and the parachutes under our sweaters. MacDougall dropped down to a low altitude and flew right over the heads of the excited crowd. While 'Spider' and I were blithely tossing out footballs and waving to the fans, the engine suddenly started to lose power. Fighting to keep the ship from barreling into the grandstand. 'Bon' frantically signaled us to pile in off the wing, which we understandably did in nothing flat. With our weight off the wing, 'Bon' was then able to get the ship high enough to miss the grandstand, but as it was we missed those top seats just by mere inches! If those people could see our features, as the newspaper item said, well we could reciprocate by saying we could see THEIRS, too! By then we knew the radiator had busted from the hot water spraying in all directions and flooding the spark plugs. We landed safely in a vacant lot a few blocks from the Coliseum.

"It sort of stuck in our craws that the paper labeled us 'flying fools' because the utmost precautions were exercised at all times for the safety of the public—how the hell did we know the radiator was going to bust?

"Anyway, after we landed we decided since we were that close, we'd just as well see the second half of the game, so we parked and tied down our ship and did just that!

"Remember those scenes in the old-time movies where two fellows would be fighting desperately in an airplane, then jump or fall out, and continue fighting on the way down, both hanging in a 'chute? Well, Art Goebel and I were sent up to Muroc Dry Lake one day to make one of those shots. Now at that high altitude, about 5,000-feet, our 'Jenny,' which at sea level could only climb to 7000-feet, could reach only 1500-feet over the lake bed. My 'opponent' was a dummy made up to look like a real man. Art crammed the 'Jenny' to its very limit, and I got the signal to go out on the wing and strap the dummy to my belt. I had a heck of a tussle with him in that wind but finally got him attached, and then wrestled him back to the trailing edge of the wing. By then he was dangling between my legs, a bad position for the jump. I was tugging and tugging, trying to pull him back on the wing when I happened to look at Goebel in the cockpit. He was waving frantically for me to jump. Then I saw the reason—we were only 200-feet from the ground! With the combined weight of both me and that stupid dummy on the wing in that thin air, Art was losing altitude so fast he was about to crash.

"Needless to say, I bailed! In the few seconds it took for the dummy and me to arrive on terra firma, I not only had no time to fight with him, but the 'chute had no time to completely open and I landed with a bang, astraddle the dummy, right in the middle of a cactus patch! It took all the rest of that day and most of the night to pull those cactus needles out of my anatomy an 'ouch!' at a time!

"We made that shot later at sea level. You may have seen it in the movies—a realistic picture of two men fighting desperately while suspended from a parachute—that blasted dummy and I!

"In conclusion I've got to tell you that us 'Cats' were especially proud of one of our pilots, Paul Richter (who with Jack Frye was later to organize TWA) the day he competed with "Swede" Olsen in the dead-stick landing contest. Now a 'dead-stick' landing is when a ship comes in for a landing with a dead motor and the propeller standing still. That's a dead-stick. A dead-stick landing contest is when a group of pilots compete by seeing who can land closest to a chalk mark on the field with a dead motor. Now 'Swede' was supposed to be the best in the business when it came to dead-stick landing events, and I guess Paul was determined to beat him and uphold the honor of the 'Black Cats.' Taking off, Paul went to 5000-feet and cut his motor, then looped, spun and rolled his ship until he was breathtakingly close to the ground. Side-slipping toward the white chalk line on the field, Paul saw that he was going to overshoot his mark. He was headed straight for the grandstand packed with people. His only out was to pull up into a loop. Missing the grandstand by a hair, Paul finished his loop and landed his ship, upside down, exactly on the line! While the audience gasped, he climbed out of the ship, dusted the chalk of the finish line off the top wing, and said nonchalantly, 'Well, that's that!' Of course, the rules of the contest specified the landing was supposed to be right side up!"

Some of the "cats"...back row standing from left to right: Al Johnson, Bon MacDougal, Art Goebel, Ken "Fronty" Nichols and Paul Richter. Front row (kneeling) left to right: Herd McClellan, Samm Greenwall and "Spider" Matlock.

In the foregoing anecdotes, Ken 'Fronty' Nichols has described a few of the '"Cat's" capers that didn't pan out precisely as planned, but then motors weren't nearly as reliable in those days as they are now and unforseen things were bound to happen. Suffice it to say that, during their "nine lives," the "13 Black Cats" put on hundreds of amazing performances utilizing every safety precaution in the book for the protection of, both, performers and public. And most of their acts were—you've guessed it!—impossible!

And, now, in bringing this book to a close, may we remind you that yesterday is today's memory and tomorrow is today's dream!

INDEX

FOREWARD - Page 5

PREFACE - " 7

A DOUBLE DILLY BY DOOLITTLE AND COLBY GOES INTO THE RED - - - " 9
    Jimmy Doolittle sums it up nicely while Tom Colby winds up with
    his bean in a paint can!

WILEY POST TAKES A DIVE - " 12
    Wiley makes a splash at Dawson and Will Rogers pegs the trouble!

A CASE OF "LEND ME YOUR EARS"! - - - - - - - - - - - - - - - - - - " 14
    A "corny" saga starring Amelia Earhart, W. B. Kinner and
    Neta Snook.

NO BOUQUETS FOR WILSON - " 17
    Wherein Claude R. Wilson demonstrates how NOT to become
    a sweet pea picker.

INTERNATIONAL INCIDENT - " 19
    Lts. "Put" Storres, "Woody" Woodring and W. L. Cornelius,
    with accomplices Bob Lautt and Vince Barnett, take a bus-
    literally!

FORSSTROM'S FLYING FOLLIES - " 21
    Carl Forsstrom makes a smashing success of his flying career.

BAIL, BOWMAN, BAIL! - " 26
    Ray R. Bowman finds the Pacific is more than a shoeful!

ANYTHING TO OBLIGE - " 28
    It was hardly Tex Rankin's plan to star in a scoop!

DID ANYONE HERE SEE JENSEN? - - - - - - - - - - - - - - - - - - - " 29
    For a pair of Goggles, Martin Jensen has the last giggle!

IT WAS ALL DOWN HILL - - BUT . . ! - - - - - - - - - - - - - - - - - " 33
    Karl Voelter and Dale (Red) Jackson find themselves in a
    "hill of note!"

THE SEAT OF THE TROUBLE - " 38
    "Ham" Lee dredges a good one from out of his past!

HOT PANTS AT THE LANGMACK HOMESTEAD - - - - - - - - - - - - - " 41
    Dave Langmack discovers too late the advantages of a
    two pants suit!

A HELLUVA WAY TO RUN AN AIRLINE! - - - - - - - - - - - - - - - - " " 43
    Johnnie Guglielmetti's passenger felt cheated, but we don't
    know why - HE had the parachute!

NO CONSIDERATION WHATEVER! - - - - - - - - - - - - - - - - - - - " 45
    Bob Lloyd learns that pilots should always have their
    forced landings on bus lines.

ONLY INVERTED, TAIL-FIRST, DEAD-STICK NIGHT LANDING IN HISTORY   Page 46
    Other than that, Charlie Langmack made a good landing!

THIS FLIGHT WAS A CORKER! --------------------------- " 48
    It took little effort for Walt Bohrer and Hans Mirow to "beer up"
    under the strain of this flight!

BASIL ROWE AND THE "SUGAR CANE PATROL"! ---------------- " 50
    When all else fails - eat the ouangas! That's what Basil Rowe
    and Bill Wade did!

FLYING - ALASKAN STYLE! ----------------------------- " 55
    Where anything can happen to bush pilots like, for instance,
    Archie Ferguson and Oscar Winchell.

RUSS MORGAN OR FREDDIE MARTIN? ---------------------- " 57
    The trouble was that Lloyd Jarman wasn't "tuned in" on
    this flight.

THE OLD LAMPLIGHTER! ------------------------------ " 59
    It's things like this that make Ray Tremblay trembly!

"AH, SO! NOW TO HITCH-HIKING!" --------------------- " 61
    All we can say for Norm Helwig here is "Rotten Ruck!"

ANY OLD SPUD IN A STORM! --------------------------- " 64
    Well, at least Bob Byers wound up with a MASHED potato!

AIR-DALE OR SKY-TERRIER? --------------------------- " 65
    For bush pilot Dean Goodwin, this flight was anything but
    a "howling" success."

"TAXI!! - OR ANYTHING!" --------------------------- " 66
    Dean Goodwin discovers that keen competitors aren't
    always too sharp!

EAGLES AND AIRPLANES DON'T MIX! -------------------- " 67
    As far as Goodwin was concerned, this hop was for the
    birds!

CANINE CLOWN OF THE NORTH! ------------------------- " 68
    With a friend like "Whitey", who needs enemies?
    Elbert E. (Al) Parmenter and Ruth Parmenter.

AND THEN THERE WAS "BOOSTER!" ---------------------- " 71
    Barnstorming with Charlie Lindbergh and "Shorty" Lynch
    was anything but a dog's life for this pooch.

GRAHAM "GROUNDS" HIS FOGGIA FARMAN! ---------------- " 73
    "Red" Graham discovered a gal can put a guy in the hole
    even if she was a half world away!

WATCH OUT - OR TIME FLIES! ------------------------- " 75
    When Freddie Lund met this farmer, he knew his "time"
    had come.

ANYONE FOR SWAHILI? -------------------------------- " 78
    Mountains they could get over, but not the language
    barrier! Bob Ceniceros, Jimmy Doolittle, Tex Rankin

GENTRY FINDS HER PLACE -------------------------------- Page 81
    Viola Gentry and Bill Ulrich find there are times when
    aviation really stinks!

HOW ABOUT THE CROWS? ----------------------------------- " 85
    Maybe Mel Hodgdon and Ted Hedlund should have Souix'd
    someone!

CERTAINLY NOT FLY-LIKUM-EAGLE! ------------------------- " 87
    Big Injuns alway did weigh on Tex Rankin - and that's
    not to be made light of!

IN PLOW LANGUAGE, "S" IS FOR SCRAM! -------------------- " 88
    Charlie McAllister and "Buss" Schlosstein get the "Indian sign"
    in a big way!

THE DAY IT RAINED LUMBER IN BURBANK -------------------- " 90
    The time test pilot Tony LeVier came "lumbering in"
    with a Ventura bomber!

UNENDURABLE ENDURANCE! --------------------------------- " 94
    It was beginning to look doubtful if Ann Bohrer could
    get down, even if she ran out of gas!

IS THIS PILOT NECESSARY? ------------------------------- " 101
    Lt. Knud von Clauson-Kaas, W. G. Mead, Harold Bromley
    and Art Jenks all thought the other guy was flying!

SPEAK PLAINER, PLEASE! --------------------------------- " 103
    Maybe Dick Depew and Charlie McAllister should've
    used cue cards!

THE COW SHOULD'VE BLOWN HER HORN! ---------------------- " 104
    This critter horned in on Joe Palmer's landing pattern!

GHOST RIDER IN THE SKY! -------------------------------- " 105
    If Charlie McAllister had been a shade lower, he'd
    have REALLY spooked 'em!

THE CURRENT PROBLEMS OF A "BALD EAGLE!" ---------------- " 107
    Thanks to Art Whitaker, this line was busy!

TALLMAN -- UNFURLED! ----------------------------------- " 109
    Much to his chagrin, Frank Tallman created quite a
    flap on this flight!

"IRON HAT", BARLINGS AND BUZZARDS ---------------------- " 112
    "Iron Hat" Johnson's li'l ol' Barling wasn't about to
    become gastronomic fare for razorbacks or buzzards!

THE "13 HOLLYWOOD BLACK CATS" -------------------------- " 116
    Some catty capers by Ken (Fronty) Nichols, "Bon"
    MacDougall, "Spider" Matlock, Art Goebel, Paul Richter
    and others.